Naomi Norman
Educational Consultant

ISBN: 0 563 47431 9
Published by BBC Educational Publishing
First published 2000
Reprinted in 2000

Designed by Cathy May (Endangered Species)
Illustrated by Kathy Baxendale
Reproduced and printed in Great Britain by Sterling

Contents

Bitesize Maths

Number

Algebra

Shape and space

Measures

Handling data

Mental maths

Introduction

KS3
BITESIZEMaths: this book!
ISBN: 0 563 47431 9

TV programmes
also available on video
Video 1
ISBN: 0 563 47434 3
Video 2
ISBN: 0 563 47442 4

The website:
www.bbc.co.uk/education/ks3bitesize/
to get more practice in those areas that you find difficult

About Key Stage 3 BITESIZE Revision

Key Stage 3 BITESIZE Revision will help you do your best in your Key Stage 3 National Tests. This maths book will help you to work towards papers covering Levels 4–6. Don't forget to use your notes from school and any textbooks you have been using.

There are six sections to this book, which are divided into units. Each unit covers a maths topic that you will have been taught during Key Stage 3 (Years 7, 8 and 9) and has an explanation of the main ideas in the topic, with "Remember!" hints and example questions. There are also practice questions. The answers to the example questions and practice questions are given at the back of the book.

Using the TV programmes and website

The TV programmes are broadcast during the night, so you will have to record them on video. You can also buy KS3 BITESIZE videos. Using the videos, you can go over the bits you need to as often as you like. The TV programmes for Units 1 to 4 are Spring 2000, and for Units 5 and 6 are Spring 2001.

The TV programme and website have cross-references to show where the book has more information. The book has TV and website symbols to show you where there is more information.

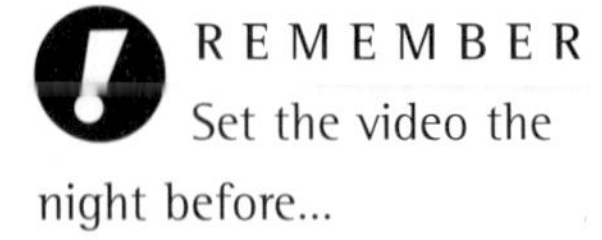

REMEMBER Set the video the night before...

Key Stage 3 maths

The National Curriculum programme of study for Key Stage 3 mathematics is divided into four Attainment Targets. These are:

Ma1 Using and Applying Mathematics
Ma2 Number and Algebra
Ma3 Shape, Space and Measures
Ma4 Handling Data

The maths tests

The Key Stage 3 tests in maths contain four different tiers of test papers. Everybody has to take Papers 1 and 2 in one of the four tiers. Each test paper lasts one hour. For Paper 1, you are not allowed to use a calculator. For Paper 2, you are allowed to use a calculator. Each mental arithmetic test is a taped test of 30 questions, with a running time of approximately 20 minutes. Tests A and B are for the higher tiers, and Test C is for the lower tier.

KEY TO SYMBOLS
TV A link to the video

Using this book to revise

- Plan your revision carefully. You'll take your Key Stage 3 maths tests in early May. Start your revision at least two months before you take your tests to be sure that you have enough time to cover all of the sections in this book. It is no good leaving your revision until the last moment.
- Breaking the subject up into BITESIZE chunks is the best way to learn. That is why this book is divided into six sections that comprise small units. Decide how many units you must cover each week to get it all done before your tests. Use the Contents page to plan your revision timetable. On your timetable, write the date when you plan to work on each unit.
- Plan the revision of all your test subjects at the same time. Make sure you build breaks and time to relax into your plan; and organise your revision around sports, hobbies and your favourite TV programmes!

REMEMBER Have this BITESIZE maths book with you as you watch the videos. Do not try to watch a whole tape at once – watch one BITE and then work through that unit in the book.

Revision tips

You revise best when you are actively doing something.

- List the important words and ideas as you work through each unit.
- After each unit, close the book and list the key facts and ideas.
- For the examples, 'Look, cover, work out' and then check.
- Write a set of summary notes for each unit.
- Test yourself by writing a mathematical question on one side of a piece of card and the working out on the other.
- Revise with a friend – use the flash cards you made to test each other.
- Record notes on audio tape, then make notes as you listen to yourself.
- Make up revision cards to fit in a pocket; test yourself when you have a moment.

There are more tips for preparing for the mental maths paper on page 87.

In the test: make sure you gain the marks

The marker can only judge how good your answer is by what you write down.

- Read the question carefully before you start to write the answer.
- Make sure you follow all of the instructions in the question.
- When you choose a number from a list, use only the numbers provided.
- Write your answer clearly and precisely.
- If you make a mistake, ensure that your first answer is crossed out and that it is clear which answer you want the marker to read.
- Show the steps in your calculations as well as the answer.
- Draw or add to diagrams using a pencil and ruler.
- Do not tick any more boxes than you are asked to in the question.

With careful, planned revision you will answer the test questions confidently and gain all the marks you deserve!

THE ON-LINE SERVICE You can find extra support, tips and answers to your exam queries on the BITESIZE internet site. The address is http://www.bbc.co.uk/education/revision

  Whole numbers

In this unit you will revise:

- place value
- multiplying by 10, 100 and 1000
- dividing by 10, 100 and 1000
- multiplying and dividing by multiples of 10, 100 and 1000

Place value

0, 1, 2, 3, 4, 5, 6, 7, 8, 9 are called **digits**.

The **place** of each digit in a number tells you its value.

Read this number out loud: two thousand, nine hundred and thirty-seven.

- Write two thousand, nine hundred and thirty-seven in digits.

The place of the right-hand digit, 7, tells you there are 7 units (seven)

The place of the next digit along, 3, tells you there are 3 tens (thirty)

The place of the next digit along, 9, tells you there are 9 hundreds (nine hundred)

The place of the next digit along, 2, tells you there are 2 thousands (two thousand)

	thousands		hundreds		tens		units
You can write this in columns:	2		9		3		7
Or you can write this as a sum:	2000	+	900	+	30	+	7

2000 is 2 thousands, 0 hundreds, 0 tens, 0 units

You can also describe 2000 in terms of units, tens, hundreds.

2000 is 2000 units

2000 is 200 tens, 0 units or 2000 is 200 tens

2000 is 20 hundreds, 0 tens, 0 units or 2000 is 20 hundreds

- Fill in the gaps to describe 900 in three ways.

900 is ...... units 900 is ...... tens 900 is ...... hundreds

REMEMBER When you multiply a number by 10, 100, 1000 your answer will be **bigger** than the number you started with.

Multiplying by 10, 100 and 1000

To multiply by 10:

move each digit one place to the left and put one zero (0) on the end

	thousands	hundreds	tens	units
			2	3
23 x 10 = 230		2	3	0

To multiply by 100:

move each digit two places to the left and put two zeros (00) on the end

	thousands	hundreds	tens	units
			4	1
41 x 100 = 4100	4	1	0	0

To multiply by 1000:

move each digit three places to the left and put three zeros (000) on the end

	thousands	hundreds	tens	units
				7
7 x 1000 = 7000	7	0	0	0

REMEMBER When you divide a number by 10, 100, 1000 your answer will be **smaller** than the number you started with.

Dividing by 10, 100 and 1000

To divide by 10:

move each digit one place to the right and cross one zero (0) off the end

	thousands	hundreds	tens	units
		6	5	0
650 ÷ 10 = 65			6	5

To divide by 100:

move each digit two places to the right and cross two zeros (00) off the end

	thousands	hundreds	tens	units
	9	7	0	0
9700 ÷ 100 = 97			9	7

To divide by 1000:

move each digit three places to the right and cross three zeros off the end

	thousands	hundreds	tens	units
8000 =	8	0	0	0
8000 ÷ 1000 = 8				8

Multiplying and dividing by multiples of 10, 100 and 1000

To multiply by multiples of tens, hundreds or thousands:
e.g. 40 x 300 = 4 tens x 3 hundreds = (4 x 10) x (3 x 100)

- multiply the 4 by the 3 — 4 x 3 = 12
- multiply the 10 by the 100 — 10 x 100 = 1000
- now multiply your answers together — 12 x 1000 = 12 000

- Write the answer to 2000 x 70 =

To divide by multiples of tens, hundreds or thousands:
e.g. 6000 ÷ 20 = 6 thousands ÷ 2 tens

- divide the number by 10, 100 or 1000 — 6000 ÷ 10 = 600
- divide the result by the number of tens, hundreds or thousands — 600 ÷ 2 = 300

- Write the answer to 900 ÷ 30 =

REMEMBER When a whole number is multiplied by another whole number, the answer is a multiple of both numbers. So 3 x 10 = 30 and 30 is a multiple of 3 and 10. Other multiples of 3 are 3, 6, 9...

Practice questions

1) Look at these seven number cards.

9	1	3	4	2	8	0

a) Amy picks four cards and makes the number 1302. Use the cards to make a four-digit number smaller than 1302.

b) Richard picks three cards to make the biggest three-digit number possible. Which three cards does Richard pick?

c) Which extra card does Richard need to pick to make his number 10 times bigger?

2) Fill in each gap with a number so that all the calculations below give the answer 400.

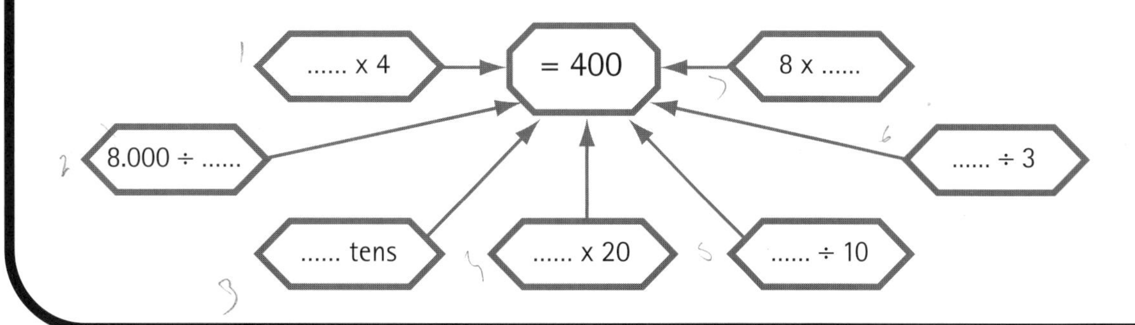

In this unit you will revise:

- understanding decimals
- rounding decimals
- reading and writing money
- adding and subtracting decimals
- multiplying and dividing by 10 and 100
- multiplying and dividing by single-digit whole numbers

Understanding decimals

REMEMBER Where you write the numbers shows their place value.

The decimal point (.) in a number shows you where the whole number stops and the decimal fraction begins. For example, fifty-eight point three one is written 58.31. The whole number is 58, the decimal fraction is .31.

You can write this in columns:	hundreds	tens	units	.	tenths	hundredths
		5	8	.	3	1

Or you can write this as a sum: $50 + 8 + \frac{3}{10} + \frac{1}{100}$

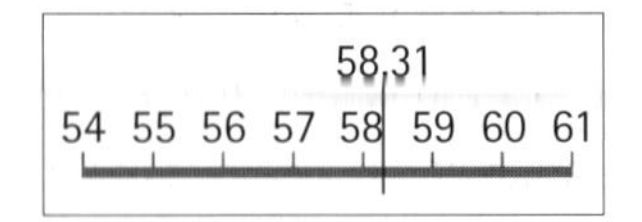

58.31
57.9 58 58.1 58.2 58.3 58.4

58.31 appears on a number line between the whole numbers 58 and 59.
In this number line you are adding units each time.

Think of a number line where you could mark 58.31 more accurately.
In this number line you are adding tenths or 0.1 each time.

The next number after 57.9 is 58. It is **not** 57.10
57.10 = 5 tens + 7 units + 1 tenth + 0 hundredths
57.9 is bigger than 57.8 because $\frac{9}{10}$ is bigger than $\frac{8}{10}$.

Rounding decimals

REMEMBER To round a decimal number, look at the digit in the place after the one to which you are rounding. If it is 0, 1, 2, 3 or 4, round off. If it is 5, 6, 7, 8 or 9, round up to the next number.

Decimals can be rounded to one, two, three or more places.

On the number line, 58.31 is closer to 58.3 than 58.4. So 58.31 rounded to 1 decimal place (1 d.p.), that is one digit after the decimal point, is 58.3.

To round 58.38 to 1 d.p. decide if 58.38 is closer to 58.3 or 58.4. As 58.38 is closer to 58.4, 58.38 is rounded up to 58.4 to 1 d.p. To round 58.35 to 1 d.p. notice that 58.35 lies halfway between 58.3 and 58.4. Numbers that end in 5 are always rounded up, so 58.35 is 58.4 to 1 d.p.

Reading and writing money

Decimals are used in money. When you use a calculator to calculate money the display will not show a zero at the end of the decimal.

£2.50 shows up as 2.5 because £0.50 is 50p, which is 0.5 of £1.

£0.40 shows up as 0.4 because £0.40 is 40p, not 4p, and 0.40 is 4 tenths of £1.

Petrol saver: **69.2p**

£0.45

For Sale £699.99

- How would you write 4p as a decimal of £1?
- How would you write the figures for this cheque?

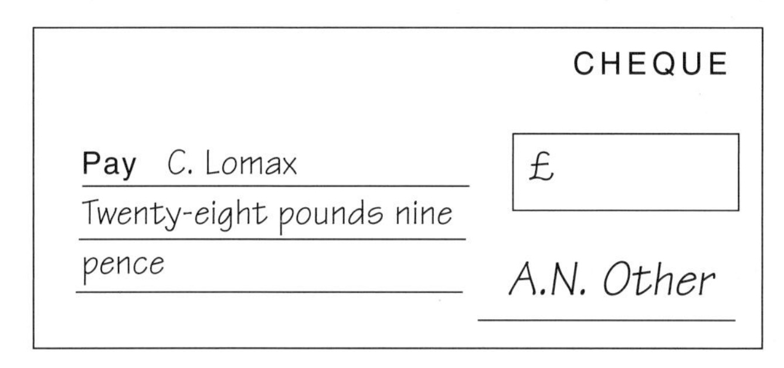

> **REMEMBER** You can put as many zeros at the end of decimals as you want – it won't change the value. They just mean 0 tenths, 0 hundredths, 0 thousandths and so on.

Adding and subtracting decimals

To add or subtract decimals, line up the decimal points so that hundreds, tens, units, tenths... are underneath each other.

Examples

19.17 – 6.42	19 . 17	125 – 81.34	125 . 00
	– 6 . 42		– 81 . 34
	12 . 75		43 . 66

- Work out 2.43 + 34.732 + 125.

> **REMEMBER** For an amount in pounds, use the £ sign: £0.24 or £1.32. For an amount in pence, use the p sign: 24p or 132p

> **REMEMBER** You can write a decimal point after whole numbers and put zeros on the end to help you line the numbers up.

Multiplying and dividing by 10 and 100

To multiply a decimal by 10:
move each digit one place to the left

For example, 46.9 can be written in columns, like this.

	hundreds	tens	units	.	tenths	hundredths
		4	6	.	9	
and 46.9 x 10 = 469	4	6	9			

REMEMBER When multiplying decimals by 10, 100 or 1000 your answer will be **bigger** than the decimal you started with.

To multiply a decimal by 100:
move each digit two places to the left.

For example, 2.7 can be written in columns, like this.

	hundreds	tens	units	.	tenths	hundredths
			2	.	7	
and 2.7 x 100 = 270	2	7	0			

- Write the answers to:

a) 3.6 x 10 = b) 63.941 x 100 =

REMEMBER When dividing decimals by 10, 100, 1000 your answer will be smaller than the decimal you started with.

To divide a decimal by 10:
move each digit one place to the right.

For example, 7.4 can be written in columns, like this.

	hundreds	tens	units	.	tenths	hundredths
			7	.	4	
and 7.4 ÷ 10 = 0.74			0	.	7	4

To divide a decimal by 100:
move each digit two places to the right.

For example, 106 can be written in columns, like this.

	hundreds	tens	units	.	tenths	hundredths
	1	0	6	.		
and 106 ÷ 100 = 1.06			1	.	0	6

- Write the answers to:

a) 32.9 ÷ 10 = b) 51.14 ÷ 100 =

Multiplying and dividing by single-digit whole numbers

To multiply and divide decimals by single-digit whole numbers line up the decimal point in the answer with the decimal point in the question.

Examples

45.61 x 3

```
  45.61
    x 7
 ------
 319.27
```

21.18 ÷ 3

```
   7.06
  ------
3)21.18
```

- Write the answers to: a) 5.613 x 4 = b) 632.4 ÷ 6 =

Practice questions

1)

100	9.10	1.03	9.11	9.09	9.8	9.1	10.2	10.3

Use these number cards to fill in the gaps in the calculations.

a) 5.10 x 2 = ...

b) 11.72 – 2.62 = ...

c) ... x 100 = 103

d) 9.10 is bigger than ...

e) 9.1 is the same as ...

f) 103 ÷ 10 = ...

g) 9.9 x ... = 990

h) $4\overline{)39.2}$ = ...

i) 5 + 4.11 = ...

2) Write the following money amounts:

a) one pound, three pence

b) 853p in pounds

c) ten pounds, thirty pence

d) 1014p in pounds.

Fractions

In this unit you will revise:

- understanding fractions
- finding equivalent fractions
- adding and subtracting fractions
- converting between fractions and decimals
- calculating fractions of whole numbers

REMEMBER When you think of fractions, imagine a picture. You might think of $\frac{1}{2}$ as  or $\frac{3}{4}$ as 

Understanding fractions

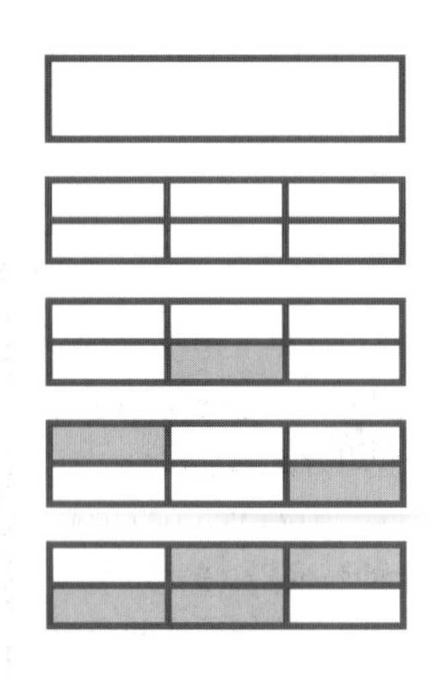

Fractions are made up of equal parts of a whole.

This rectangle is one whole. It you split it into six equal parts, each part is $\frac{1}{6}$

If any one part is shaded, then $\frac{1}{6}$ is shaded and $\frac{5}{6}$ is not shaded.

If any two parts are shaded, then $\frac{2}{6}$ is shaded and $\frac{4}{6}$ is not shaded.

If any four parts are shaded, then $\frac{4}{6}$ is shaded and $\frac{2}{6}$ is not shaded.

- In this diagram, what fraction is shaded? What fraction is not shaded?

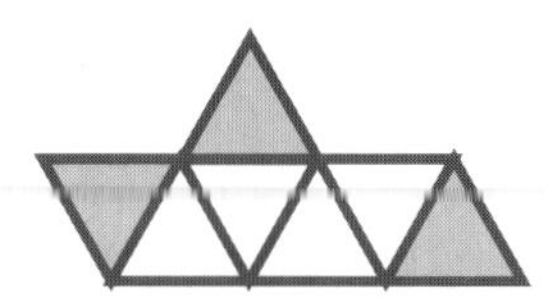

$\frac{5}{6}$ numerator / denominator

The number below the line (the denominator) shows how many equal parts there are in the whole. The number above the line (the numerator) shows how many of these equal parts there are in the fraction. So $^5/_6$ means the whole is split into six equal parts (sixths) and there are five of them in the fraction.

Finding equivalent fractions

$\frac{1}{2}$ of this square has been shaded 

$\frac{4}{8}$ of this square had been shaded 

REMEMBER If you multiply or divide the top and bottom of a fraction by the same number, its value does not change.

You can see that the same area has been shaded in each square.

So $\frac{1}{2} = \frac{4}{8}$

The top and bottom of $\frac{1}{2}$ have both been multiplied by 4 to make $\frac{4}{8}$.

$\frac{6}{8} = \frac{6 \div 2}{8 \div 2} = \frac{3}{4}$

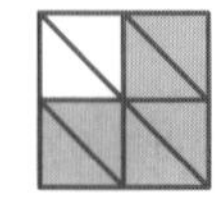 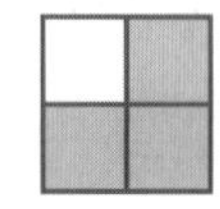

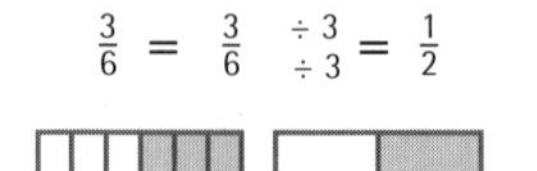

$\frac{3}{6} = \frac{3 \div 3}{6 \div 3} = \frac{1}{2}$

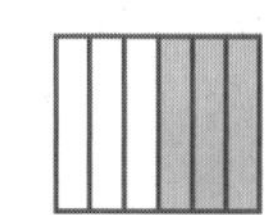 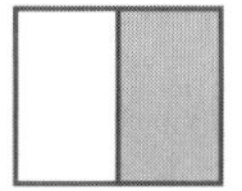

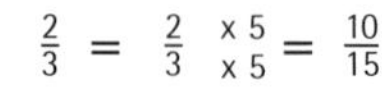

$\frac{2}{3} = \frac{2 \times 5}{3 \times 5} = \frac{10}{15}$

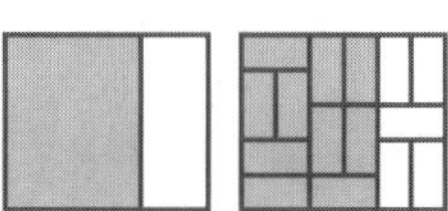

Adding and subtracting fractions

You can only add or subtract fractions if they have the same denominator. For example, quarters can only be added to or subtracted from quarters, ninths can only be added to or subtracted from ninths.

If you need to add or subtract fractions with different denominators, you must find an equivalent fraction for one or both.

REMEMBER Fractions must have the same denominator before they can be added or subtracted.

Example Find $\frac{5}{6} - \frac{2}{9}$

$\frac{5}{6}$ has denominator 6 and $\frac{2}{9}$ has denominator 9.

You cannot find an equivalent fraction to $\frac{5}{6}$ with a denominator of 9, nor an equivalent fraction to $\frac{2}{9}$ with a denominator of 6.

You need to find equivalent fractions to both $\frac{5}{6}$ and $\frac{2}{9}$ with a denominator that both 6 and 9 divide into 18.
For $\frac{5}{6}$, multiply top and bottom by 3 to give $\frac{15}{18}$. For $\frac{2}{9}$, multiply top and bottom by 2 to give $\frac{4}{18}$.
Then $\frac{5}{6} - \frac{2}{9} = \frac{15}{18} - \frac{4}{18} = \frac{11}{18}$

- Try this calculation: $\frac{1}{3} + \frac{3}{5}$

Converting between fractions and decimals

You can write fractions in columns, just as you do for whole numbers.

	units	.	tenths
$\frac{1}{10}$ is one tenth =	0	.	1
$\frac{3}{10}$ is three tenths =	0	.	3

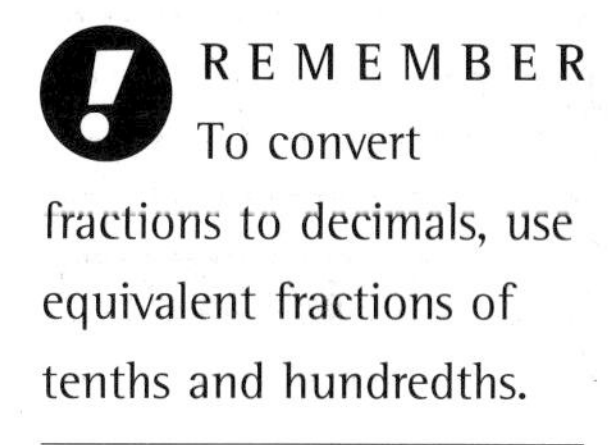

It is not so easy for the fraction $\frac{1}{2}$ because it does not have a denominator of 10. You must find an equivalent fraction with a denominator of 10, by multiplying top and bottom by 5. $\frac{1}{2} = \frac{5}{10}$

Sometimes it is not possible to find an equivalent fraction with the denominator 10. Think about $\frac{1}{4}$. The denominator 4 cannot be multiplied by a whole number to make 10. The next column in the place value chart is hundredths, so try to make an equivalent fraction with denomionator 100. You can make an equivalent fraction for $\frac{1}{4}$ with a denominator of 100 by multiplying top and bottom by 25.

	units	.	tenths	hundredths
$\frac{1}{4} = \frac{25}{100} =$	0	.	2	5

You can use a calculator to convert more difficult fractions to decimals.

To convert $\frac{1}{3}$, key in 1 ÷ 3 on your calculator. You should get the answer 0.3333333333. The 3 just keeps on repeating, for ever.

> **REMEMBER** You can convert fractions to decimals by dividing the numerator of a fraction by the denominator, on a calculator. Try 1 ÷ 4 on your calculator. You should get the answer 0.25.

Decimals that do this are called recurring decimals and are often written with a dot above the number that repeats. So $\frac{1}{3}$ is written as the decimal $0.\dot{3}$.

Calculating fractions of whole numbers

To find a fraction of a whole number:

- find one part of the fraction For example, to find $\frac{2}{5}$ of 35, find $\frac{1}{5}$ first. Divide 35 by 5.

  $\frac{1}{5}$ of 35 = 35 ÷ 5 = 7

- multiply by the number above the line (the number of parts you want) For example, to find $\frac{2}{5}$ of 35, multiply $\frac{1}{5}$ of 35 by 2.

  $\frac{2}{5}$ of 35 = 2 x $\frac{1}{5}$ of 35 = 2 x 7 = 14

- Find $\frac{2}{3}$ of 27.

Practice questions

1) Look at these five fraction cards A – E. Some show fractions represented by digits, others show fractions represented by their shaded areas.

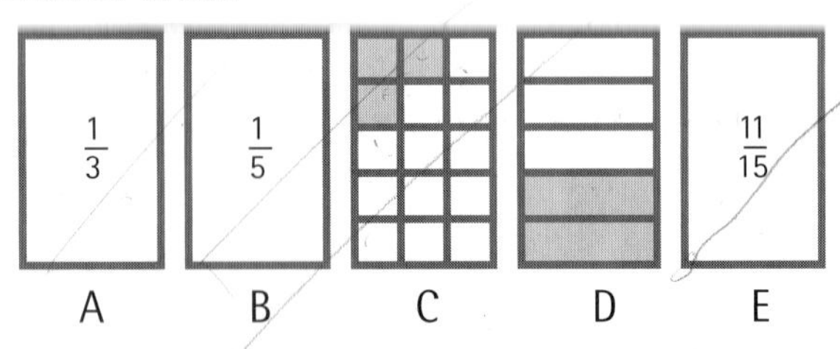

Fill in the blanks with the correct fractions from cards C and D.

a) ______ of card C is shaded

b) ______ of card D is shaded

Only one of each of the cards can be used to answer the following questions. Cross off each card as you use it.

c) Antony picks the biggest fraction of all. Which card does Antony pick?

d) Crista picks the card that is equivalent to $\frac{2}{6}$. Which card does Crista pick?

e) Ross and Louise pick the two fraction cards that are equivalent to each other. Which two cards do they pick?

f) David picks the fraction card that is twice as big as each of Ross' and Louise's cards. Which card does David pick?

2) Wendy wins £120 in a competition. She gives some of her winnings to charity.

a) Wendy donates one eighth to Children in Need. How much does Children in Need get?

b) Wendy donates two-fifths to Comic Relief. Find out how much Comic Relief gets.

c) Wendy donates five-twelfths to ChildLine. How much does ChildLine get?

d) Wendy keeps the rest of the money for herself. How much does she keep?

Percentages

In this unit you will revise:

- understanding percentages
- calculating percentages
- calculating percentage changes
- expressing one number as a percentage of another number

Understanding percentages

Per cent (written %) means 'in every hundred'.

The whole is 100%, which means '100 in every hundred'.

Percentages are fractions of a whole, expressed in hundredths.
The denominator is 100, but instead of writing it in you use the % sign.

50% is $\frac{50}{100}$ which is the same as $\frac{1}{2}$

25% is $\frac{25}{100}$ which is the same as $\frac{1}{4}$

Percentages can be written as decimals too.

50% is $\frac{50}{100}$ which is $50 \div 100 = 0.5$

25% is $\frac{25}{100}$ which is $25 \div 100 = 0.25$

Percentages, fractions and decimals are all linked together. Look at these patterns.

25% + 50% = 75%	10% + 20% = 30%
$\frac{25}{100} + \frac{50}{100} = \frac{75}{100}$	$\frac{10}{100} + \frac{20}{100} = \frac{30}{100}$
0.25 + 0.5 = 0.75	0.1 + 0.2 = 0.3

REMEMBER Percentages are like equivalent fractions, where the denominator is 100. Turn to page 14 to read about equivalent fractions.

FactZONE

Common percentages you should know:

$25\% = \frac{25}{100} = \frac{1}{4} = 0.25$

$33.3\% = \frac{1}{3} = 0.33$

$20\% = \frac{20}{100} = \frac{1}{5} = 0.2$

$30\% = \frac{30}{100} = \frac{3}{10} = 0.3$

$10\% = \frac{10}{100} = \frac{1}{10} = 0.1$

$50\% = \frac{50}{100} = \frac{1}{2} = 0.5$

$75\% = \frac{75}{100} = \frac{3}{4} = 0.75$

Calculating percentages

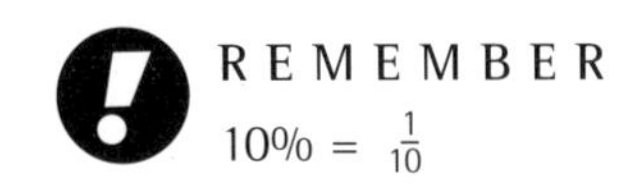

REMEMBER
$10\% = \frac{1}{10}$

It's easy to find percentages that are multiples of 10. Just find 10% of the number first, then multiply by the number of 10 per cents you need.

Example 1 Find 20% of £30 and 30% of £30

Find 10% of £30:		10% of £30 = £3
and 20% = 10% x 2	so	20% of £30 = £3 x 2 = £6
and 30% = 10% + 20%	so	30% of £30 = £3 + £6 = £9

Finding percentages that end in 5 are quite easy, too. Find 10%, then halve it to find 5%. Multiply by the number of 10%s you need, then add on 5%.

Value Added Tax (VAT) is a tax added to the price of many goods.
VAT is $17\frac{1}{2}\%$.

You can work out $17\frac{1}{2}\%$ of £40, starting with 10%. Break $17\frac{1}{2}\%$ down into bits that you calculate. $17\frac{1}{2}\% = 10\% + 5\% + 2\frac{1}{2}\%$, and $2\frac{1}{2}$ is half of 5.

REMEMBER
% means ÷ 100
'of' means x (multiply)
Read: 35% of £22
Translate: $\frac{35}{100}$ x 22
Work out: 35 x 22 ÷ 100

10% of £40 is £4   5% of £40 is £2 $2\frac{1}{2}\%$ of £40 is £1

so $17\frac{1}{2}\%$ of £40 = 10% + 5% + $2\frac{1}{2}\%$ = £4 + £2 + £1 = £7

Example 2 Finding a more difficult percentage

Finding 23% of 50 isn't too difficult, if you remember that 23% is 23/100.

- Multiply the number by the percentage   23 x 50 = 1150
- Now divide by 100   1150 ÷ 100 = 11.5

- a)Find 56.5% of £30   b) Find 9% of 124 grams

Calculating a percentage change

To calculate a percentage change:

- Work out the change in the amount
- Write the change in the amount as a fraction of the original amount
- Multiply the fraction by 100 to change it to a percentage

Example

Kamal bought the parts to build himself a computer. They cost him £680. He sold the computer for £986. What was his percentage profit?

Change in amount: £986 – £680 = £306

Change in amount as a fraction of the original amount: $\frac{306}{680}$

The fraction as a percentage: $\frac{306}{680}$ x 100 = 45%

Expressing one number as a percentage of another number

You can write numbers as percentages of other numbers, to make it easy to compare them. To express one number as a percentage of another, write the first number as a fraction of the second and then multiply the fraction by 100 to change it to a percentage.

Example

Larry's Motormart is offering sale prices on 128 of their stock of 200 cars.
Gerry's Garage is offering sale prices on 63 of their stock of 90 cars.
Which showroom is having the bigger sale?
Larry is offering 128 out of 200 cars at sale prices.
As a fraction that is $\frac{128}{200}$ so the percentage is $\frac{128}{200}$ x 100 = 64%
Gerry is offering 63 out of 90 cars at sale prices.
As a fraction that is $\frac{63}{90}$ so the percentage is $\frac{63}{90}$ x 100 = 70%

Therefore Gerry is having the bigger sale.

Practice questions

1) Look at these eight sales tags. There are four pairs of cards which both offer the same discount. List the four pairs.

A Today only $1/3$ off

B 25% SALE

C 40% savings

D $1/10$ off

E SALE 33.3% off

F

G 10% off TODAY

H

2) Carlo's dad sells Carlo's old bicycle for £60.

He agrees Carlo can have 12.5% of the £60 to spend. See how Carlo works out 12.5% of £60 in his head.

50% of £60 is £30
25% of £60 is £15
12.5% of £60 is £7.50

75% of the £60 is to go to buying Carlo a new bicycle.

10% of the £60 is to go into Carlo's savings.
2.5% is to be given to Carlo's sister.

a) Copy the working and fill in the three gaps to show how Carlo can work out 75% of £60 in his head.

50% of £60 is £30

☐ % of £60 is £ ☐

so 75% of £60 is £ ☐

b) Work out 10% of £60 to show how much money goes to Carlo's savings.

c) Work out 2.5% of £60 to show how much money Carlo's sister gets.

3) Find the following percentages. Use your calculator if you want to.

a) 62% of 40 metres

b) 37% of 280 grams

c) 94% of £88

4) There are 11 boys in a class of 20 students. What percentage are boys?

In this unit you will revise

- understanding negative numbers
- adding and subtracting negative numbers

Understanding negative numbers

REMEMBER Always look at the sign in front of a number to see if it is positive or negative.

Numbers above zero are **positive** numbers. They are written with no sign or a + sign in front of them, for example 4, +51, +7, 39, +243.

Numbers below zero are **negative** numbers. They are always written with a minus – sign in front of them, for example –3, –21, –993, –1082.

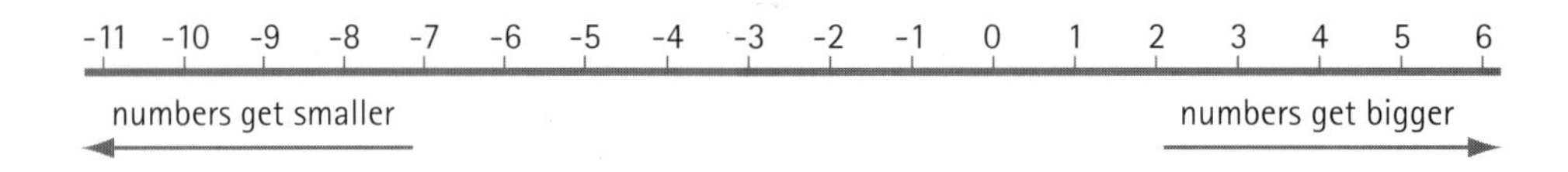

REMEMBER Count to the right ➜ for positive numbers and to the left ← for negative numbers.

Positive numbers get bigger the further you move to the right of zero.

So 3 is smaller than 5 and +6 is larger than +2.

Negative numbers get smaller the further we move to the left of zero.

So –11 is smaller than –7 and –4 is larger than –9.

- Copy these and write smaller or larger in the gaps.

a) –10 is __________ than –24 b) –183 is __________ than –83

Adding and subtracting negative numbers

To add and subtract numbers, imagine a number line with negative numbers to the left of 0 and positive numbers to the right. Always count from zero.

When dealing with positive numbers count to the right.
When dealing with negative numbers count to the left.

Example

To calculate 3 – 5 – 1

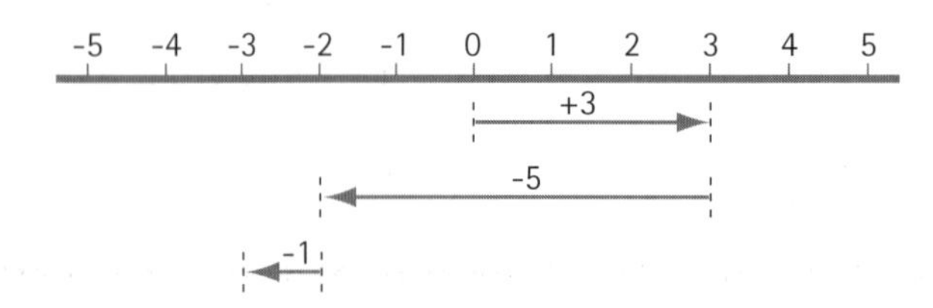

Start from zero and count +3
now count –5 (taking you to –2)
now count –1 (taking you to –3)
So 3 – 5 – 1 = –3

- Write the answer to –5 + 10 – 2 – 7

You can use a number line to help you.

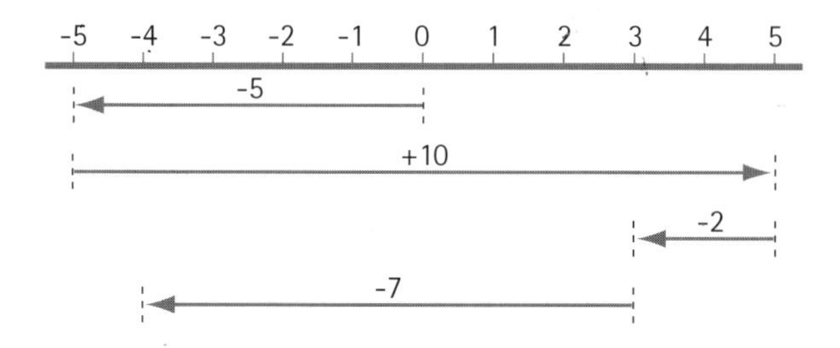

Practice questions

1) a) Put the following temperatures in order, from coldest to hottest.

–1°C, 7°C, –15°C, 0°C, 12°C, –8°C

____

b) Jack grows all sorts of plants in his greenhouse. He must keep the greenhouse temperature above –2°C otherwise his plants will die. Which of the temperatures above will kill Jack's plants?

____

2) Each of these 10 number cards fits just one of the 10 gaps below.

0	+5	-7	-4	+8	-10
+4	-20	-2	-8		

Cross off each number card as you use it to fill in the gaps.

Choose five number cards to fill in the missing numbers on the number lines.

a)

...... -15 -10 -5 0 2

b)

-6 .... 2 4 6

Now use the rest of the number cards to fill in the gaps in the calculations.

c) +2 – 10 = __________

d) -3 + 2 – 9 = __________

e) ____ – 8 = 0

f) ____ – 11 = __________

Rounding and approximating

In this unit you will revise:

- rounding to the nearest whole number
- rounding to the nearest 10, 100 or 1000
- approximating answers to calculations

REMEMBER Rounding and approximating occur in lots of questions in maths. Even if you don't have a question just on rounding and approximating, you will need to look out for questions that need you to use them in the calculations.

Rounding to the nearest whole number

- A number with a decimal part of 0.5 or more is rounded up.
- A number with a decimal part less than 0.5 is rounded down.

For example, 16.8 is rounded up to 17 because it is closer to 17 than 16 and 20.39 is rounded down to 20 because it is closer to 20 than 21

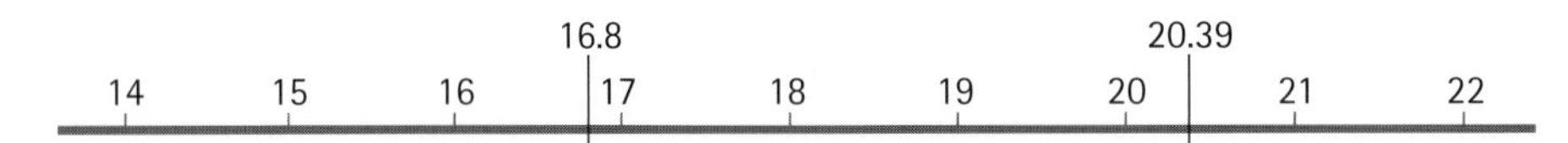

- Write each of these, rounded to the nearest whole number.

a) 12.5 b) 10.19

Rounding to the nearest 10, 100 or 1000

When you round to the nearest 10:

- a number with 5 or more units is rounded up to the next 10
- a number with less than 5 units is rounded down to the 10 below

For example, 86 (6 units) is rounded up to 90 and134 (4 units) is rounded down to 130

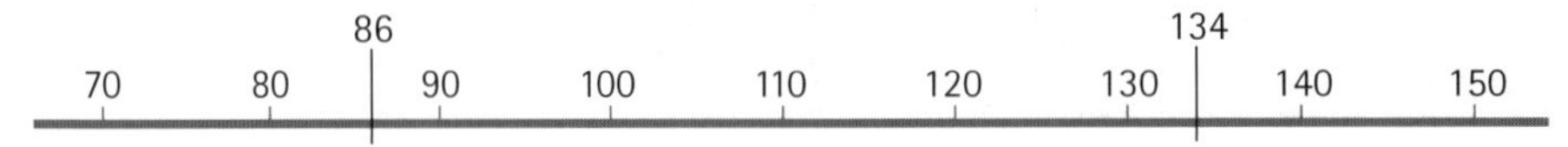

When you round to the nearest 100:

- a number with 5 or more tens is rounded up to the next 100
- a number with less than 5 tens is rounded down to the 100 below.

For example, 371 (7 tens) is rounded up to 400 and 839 (3 tens) is rounded down to 800

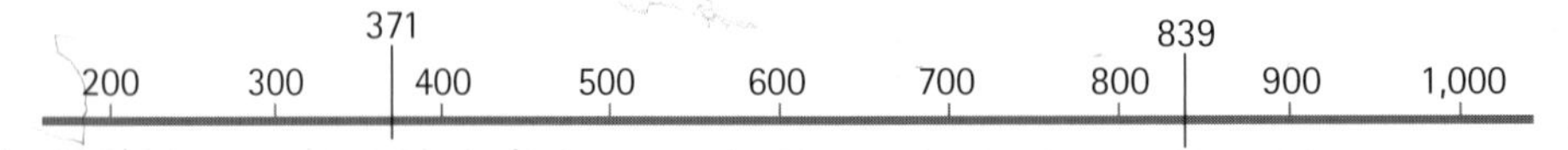

When you round to the nearest 1000:

- a number with 5 or more hundreds is rounded up to the next 1000
- a number with less than 5 hundreds is rounded down to the 1000 below.

For example, 9810 (8 hundreds) is rounded up to 10 000

9,810

4,000 5,000 6,000 7,000 8,000 9,000 10,000 11,000 12,000

a) Write 251 to the nearest hundred. b) Write 2883 to the nearest ten.

Approximating answers to calculations

- First round each number in the calculation.
- Then work out the calculation in your head.

Example

Find an approximate answer to the calculation $\dfrac{188.3 - 76.47}{1.9}$

188.3 is about 200 76.47 is about 80 1.9 is about 2

So $\dfrac{188.3 - 76.47}{1.9}$ is approximately $\dfrac{200 - 80}{2} = \dfrac{120}{2} = 60$

Practice questions

1) Look at these five calculator displays.

a) b) c) d) e)

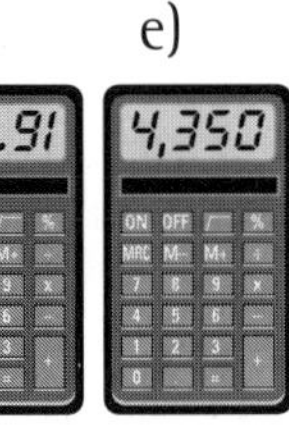

27.91 4,350

Round:

display a) to the nearest thousand.

display b) to the nearest hundred.

display c) to the nearest ten.

display d) to the nearest whole number.

display e) to the nearest hundred.

2) Find an approximate answer for the following calculations.

a) 38.7×41.07

b) $\dfrac{291.6 - 58.25}{6.2}$

3) The local post office sells novelty pens in packs of 10. One pack costs £3.99.

How many pens can you buy with a £20 note?

Unitary method and ratio

In this unit you will revise:

- understanding the unitary method
- solving money problems using the unitary method
- ratio

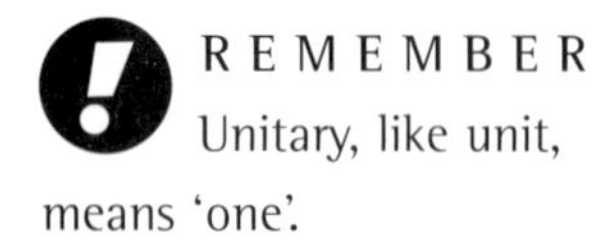
REMEMBER
Unitary, like unit, means 'one'.

Understanding the unitary method

When you know a quantity such as the weight, length or cost of several items, you can use division to find the weight, length or cost of one. When you have the weight, length or cost of one (the unit), you can use multiplication to find the weight, length or cost of any number of items. This is called the unitary method.

Example

Steph's shelf is 32 cm long, and four mugs fit on it exactly. Steph buys two more mugs. How long a shelf does she need now?

Using the unitary method: 4 mugs take 32 cm

Use division to calculate the length of 1: 1 mug takes 32 ÷ 4 = 8 cm

Use multiplication to calculate the length of 6: 6 mugs take 6 x 8 = 48 cm

- Work out how long the shelf needs to be for 8 mugs, then 11 mugs.

Solving money problems using the unitary method

The unitary method is often used to solve money problems.

Example

Jon buys a bumper pack of 12 packets of chilli crisps for £3.00. He opens the pack and eats 1 packet but doesn't like them, so he sells the rest on to Stuart, charging him for 11 packets of crisps. How much does Stuart pay?

Use division to calculate the price of 1 packet.

If 12 packets cost £3.00, 1 packet costs £3.00 ÷ 12 = £0.25

Use multiplication to calculate the price of 11 packets.

11 packets cost £0.25 x 11 = £2.75

Ratio

Ratios are another way of comparing quantities and are written as a:b.

Example 1

Nadia is making smoothies. She mixes two parts cranberry juice to three parts orange juice. This can be written as the ratio cranberry:orange, 2:3. Every 2 litres of cranberry juice requires 3 litres of orange juice. So 4 litres of cranberry juice requires 6 litres of orange juice and 6 litres of cranberry juice requires 9 litres of orange juice.

Nadia adds bananas and strawberries in the ratio 1:6

How many strawberries will Nadia need if she adds $\frac{1}{2}$ a banana?

She will need 3 strawberries, because 1 banana requires 6 strawberries.

So $\frac{1}{2}$ a banana requires 3 strawberries.

Example 2

Lou divides £30 between his granddaughters in the ratio of their ages. Abbie is 6 years old and Naomi is 9 years old, so the ratio of their ages is 6:9.

So £6 given to Abbie means £9 given to Naomi. That's £15 given in total.

£12 given to Abbie means £18 given to Naomi. That's £30 given in total.

Practice questions

1) Sharminee, Carolyn and Ginny plan to go on a windsurfing course. Sharminee writes away to find out how much it will cost and is quoted a price of £105 for the group. Then Emma decides to join in too. How much will the group price be now?

2) Miss Roberts is organising an ice-skating trip for the 28 students in her tutor group. She is quoted an admission cost for the whole tutor group of £61.60. On the day, two of her tutor group are absent. How much will the admission cost be now?

3) Mittey United football team dye their shorts and socks green in 24 pints of green dye mixture.

To make the green dye mixture they use 3 parts yellow to 5 parts blue.

a) How much yellow dye do they use?

b) How much blue dye do they use?

Simple formulae

In this unit you will revise:

- using letters for numbers
- understanding formulae
- solving simple word problems in algebra
- adding and subtracting like terms
- solving more difficult word problems in algebra

Using letters for numbers

REMEMBER It does not matter what letters you use in an algebraic term or expression, as long as you know what quantities they represent.

In algebra letters are used instead of unknown numbers. A letter may appear on its own, or multiplied by a number, or multiplied by another letter. A number or letter, or product of numbers and letters, is called a "term".

a $2b$ xy are **all terms**.

Terms may be joined by + or – to form an **algebraic expression**.

So $b + d$ is an expression. It means a number b, add another number d

$s - t$ is an expression. It means a number s, subtract another number t

Terms may be combined by x or ÷ to form expressions such as $\frac{x}{y}$ or $\frac{ab}{c}$.

Example

REMEMBER Turn to page 11 to read about problems with money.

The cost of a sandwich is p and the cost of a drink is m.

For my lunch I always have a sandwich and a drink. The cost = $p + m$.

If I order a cheese sandwich, then $p = 1.75$ (£1.75)
If I order a chicken tikka sandwich, then $p = 2$ (£2.00)
If I buy a can of fizzy drink, then $m = 0.55$ (£0.55)
If I buy a bottle of water, then $m = 0.45$ (£0.45)

On Mondays and Tuesdays I have a cheese sandwich and a bottle of water.
The cost of my lunch on Mondays and Tuesdays: $p + m = 1.75 + 0.45 =$ £2.20

On Wednesdays I have a chicken tikka sandwich and a can of fizzy drink.
The cost of my lunch on Wednesdays: $p + m = 2 + 0.55 =$ £2.55

- On Thursdays and Fridays I have a chicken tikka sandwich and a bottle of water. Write the cost of my lunch on Thursdays and Fridays.

Understanding formulae

A **formula** is a rule that can be applied to a number to give a new number.

Sometimes formulae are shown as number machines. A number is put into the machine, the rule changes it and a new number comes out.

Example 1

in → out

16 → + 9 → 25

Example 2

in → out

40 → ÷ 5 → 8

Sometimes different numbers might be put in the same number machine.

Example 3

in		out	in		out	in		out
8 →	− 6 →	2	12 →	− 6 →	6	29 →	− 6 →	23

A number is put in and that number minus 6 comes out.

You can use a letter, such as x, to represent the number that is put in.

$x \rightarrow -6 \rightarrow x - 6$

Then the number that comes out is $x - 6$.

So if $x = 8$, then $x - 6 = 2$ if $x = 12$, then $x - 6 = 6$
if $x = 29$, then $x - 6 = 23$

- Fill in the gaps for these number machines.

a) in $g \rightarrow \times 4 \rightarrow$ out □

b) in $16 \rightarrow + 11 \rightarrow$ out □

c) in $h \rightarrow \div 5 \rightarrow$ out □

Sometimes a number machine will use more than one rule.

Example 4

in → out

24 → ÷ 4 → + 7 → 13

Example 5

in → out

$n \rightarrow \times 6 \rightarrow -1 \rightarrow 6n - 1$

- Fill in the gaps for these number machines.

a) in $11 \rightarrow \times 4 \rightarrow - 10 \rightarrow$ out □

b) in $t \rightarrow \div 3 \rightarrow + 9 \rightarrow$ out □

! REMEMBER
There are several ways of solving linear equations. You may have learnt a different method at school. Perhaps you use the balance method shown on the KS3 Bitesize Maths TV programme. If you are confident with the way you know, you can skip Example 1 and Example 2 and go on to Example 3.

! REMEMBER
In algebra $g \times 4$ is usually written $4g$ and means $g + g + g + g$
$p \times 3$ is written $3p$ and means $p + p + p$
$h \div 5$ is usually written as $\frac{h}{5}$

$m \div 14$ is written $\frac{m}{14}$

Sometimes a number machine will have an operation missing.

Example 6

in out

$a \rightarrow \square \rightarrow 8a$

x 8

Example 7

in out

$b \rightarrow \square \rightarrow \square \rightarrow 4b + 9$

x 4 + 9

- Fill in the rules for these number machines.

a) in out

$p \rightarrow \square \rightarrow \frac{p}{4}$

b) in out

$r \rightarrow \square \rightarrow \square \rightarrow 5r - 11$

Solving simple word problems in algebra

You can use algebra to solve everyday problems.

In her cheesecake recipe, Grandma Renee scribbled down this instruction.

Bake in centre of oven, 320°F.

Renee's granddaughter wants to make the recipe, but her oven uses Celsius.

Renee's granddaughter finds this rule for changing degrees Fahrenheit into degrees Celsius.

subtract 32 → multiply by 5 → divide by 9

At what temperature, in °C, should Renee's granddaughter cook the cheesecake?

This problem can be written as a number machine.

320°F → − 32 → x 5 → ÷ 9 → 160°C

because $320 - 32 = 288$ $288 \times 5 = 1440$ $1440 \div 9 = 160$

REMEMBER So $4x$, $7x$ and $13x$ are like terms, but $6d$, $5e$ and $11f$ are unlike terms; 5, g and $4m$ are also unlike terms.

Adding and subtracting like terms

In algebra, **like terms** are letters, or groups of letters, that are alike. They have the same letter. You can only add or subtract like terms.

When you add or subtract like terms, it is called **simplifying**.

Example 1

$4a + 9a = 13a$

Example 2

$12c - 7c = 5c$

- Simplify these expressions.

a) $15u + 2u$

b) $v + 6v$

Example 3

$8s + 4t - 3s = 8s - 3s + 4t = 5s + 4t$

Example 4

$4p + 8q + p - 11q = 4p + p + 8q - 11q = 5p - 3q$

- Simplify these expressions.

a) $5m + 2m - 6n$ b) $13y - 2 + 2y - 11$

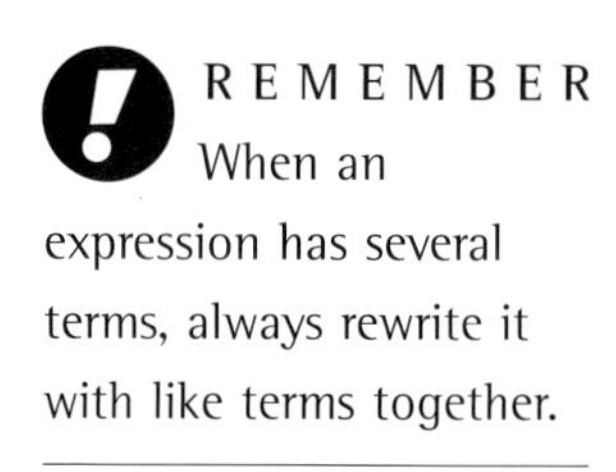

When an expression has several terms, always rewrite it with like terms together.

Solving more difficult word problems in algebra

You can use algebra to solve more difficult everyday problems. too.

Example

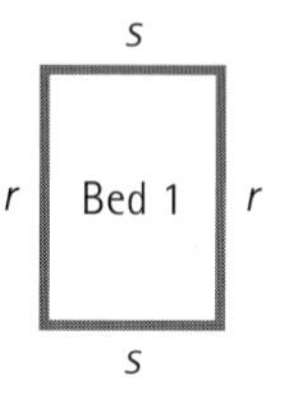

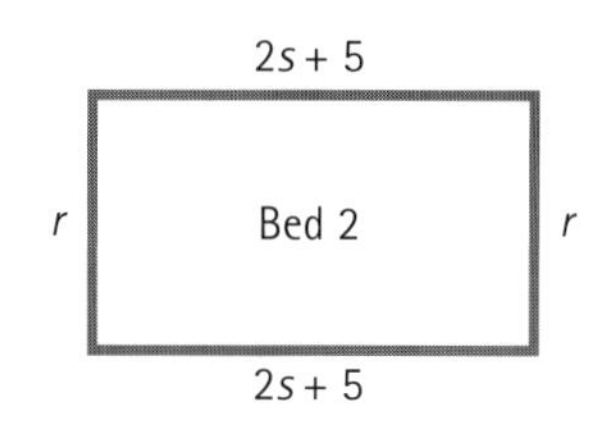

Jacqueline and Frank want to plant two flower beds in their garden, each surrounded by a miniature fence to keep their pet rabbit out!

The sizes of the flower beds depend on how many plants they can afford to buy.

They draw a plan of each flower bed.

For Bed 1 they need $r + s + r + s = 2r + 2s$ units of fencing

For Bed 2 they need $2s + 5 + r + 2s + 5 + r$ units of fencing

$= 2s + 2s + r + r + 5 + 5 = 2r + 4s + 10$ units of fencing

They decide to put the flower beds next to each other to save fencing.

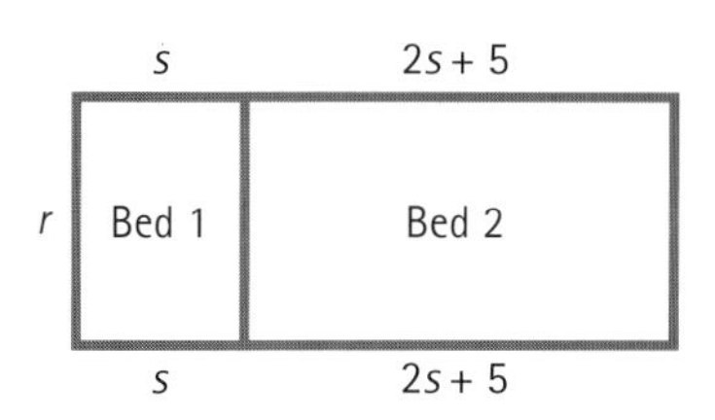

Frank says they will need $4r + 6s + 10$ units of fencing. Is Frank right?

Frank is not right.

They will need $s + 2s + 5 + r + 2s + 5 + s + r$

$= s + 2s + 2s + s + r + r + 5 + 5 = 2r + 6s + 10$ units of fencing

Practice questions

1) Copy the number machines. Fill in the gaps.

a) in out
54 → ÷ 6 → □

b) in out
5 → + □ → 13

c) in out
□ → x 4 → $4y$

d) in out
z → x 3 → − 8 → □

e) in out
12 → ÷ □ → + 1 → 5

2) each edge of this square tile is y cm long.

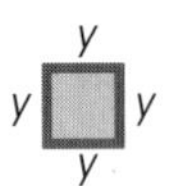

a) What is the tile's perimeter?

5 tiles are used to make the bridge shape.

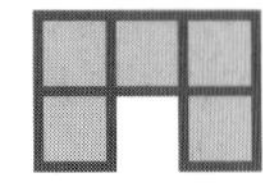

b) Write the perimeter of the bridge, in terms of y.

c) The perimeter of the bridge is 60 cm. What is y?

3) I am x years old. In my family there's Mum, Dad, my sister and our cat.

My sister is 4 years older than I am. I can write her age in terms of x as $x + 4$.

a) I am two years older than our cat. Write his age in terms of x.

b) My Dad is three times my age. Write his age in terms of x.

c) My Dad is 45. What is my age (the value of x)?

d) My Mum's age can be worked out using the following number machine.

in out
x → x 2 → + 9 → Mum's age

What is my Mum's age?

4) My sister is designing a garden patio between an old path that leads from our house into our garden, and a new shed that is yet to be built. Look carefully at her plan.

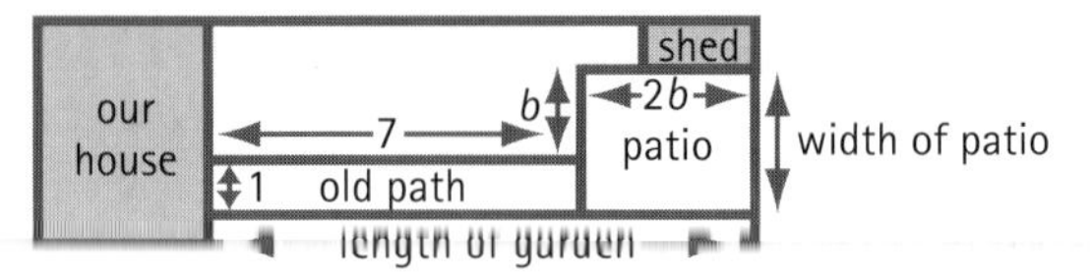

The length of the old path is 7 units.
The width of old path is 1 unit.

Look at lengths b and $2b$.

Write expressions for:

a) the length of the garden

b) the width of the patio.

Using formulae

In this unit you will revise:

- using formulae

You can use formulae to represent situations where the numbers change. You can substitute appropriate numbers for the letters in the formula.

Example 1

The cooking time for a stuffed chicken is given by the formula:

cooking time (minutes) = 40 x weight of chicken (kilograms) + 20

This is written as $t = 40w + 20$

When you buy a 2 kg chicken for Sunday lunch for four people
$t = 40 \times 2 + 20$
$t = 100$ minutes (or 1 hour 40 minutes)

When you buy a 3 kg chicken for six people
$t = 40 \times 3 + 20$
$t = 140$ minutes (or 2 hours 20 minutes)

Example 2

The bill (without VAT) for Jo's mobile 'phone is calculated using the formula:

cost, c (£) = 0.17 (£) x number of minutes, m + monthly rate, r (£)

This is written as $c = 0.17m + r$

Last November, Jo spoke for 50 minutes and her monthly rate was £18.50, so

$c = 0.17 \times 50 + 18.50 =$ £27

In December, Jo spoke for 120 minutes, but took advantage of a special deal.

> Merry Christmas to all out customers!
> Special monthly rate for December only – **£15**

So now $r =$ £15 not £18.50

So $c = 0.17 \times 120 + 15 =$ £35.40

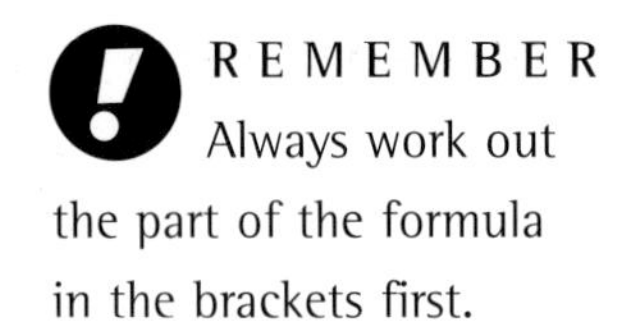

REMEMBER Always work out the part of the formula in the brackets first.

An approximate formula to convert degrees Celsius (C) to degrees Fahrenheit (F) is:

$C = (F - 30) \div 2$ Notice the brackets.

Start by working out $F - 30$. Then divide the answer by 2.

- Use this formula to write the temperatures in Celsius for these cities.

a) London: 68°F b) New York: 86°F c) Milan: 72°F

The formula for finding the area of a triangle is $\frac{1}{2}$ x base x height

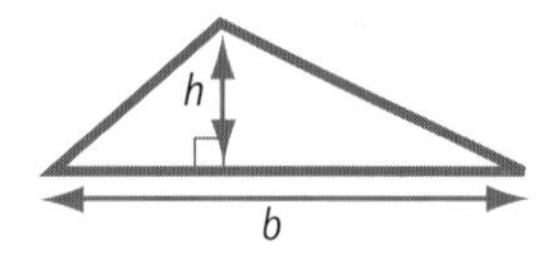

- Use this formula to find the area of this triangle.

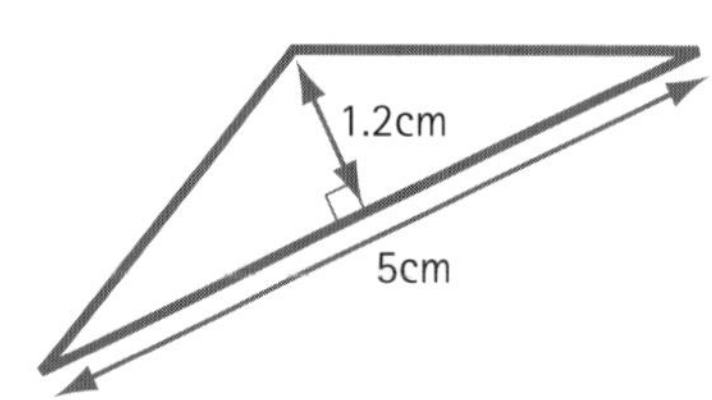

The formula for finding the volume of a cuboid is length x width x height

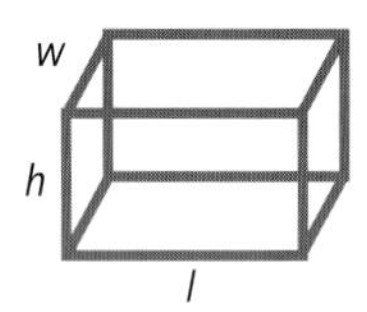

REMEMBER Turn to page 69 to read more about area and to page 72 to read about volume.

- Use this formula to find the volume of this cuboid.

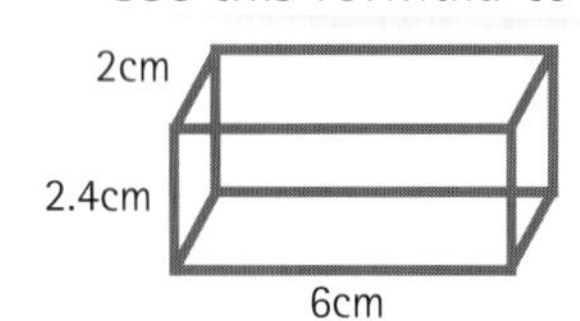

Practice questions

1) Dougie and Adam enter a 60 kilometre cycle race. The race begins at midday.

   Dougie reaches the finish line at 4pm. Adam reaches the finish line at 5pm.

   Their speed (in kilometres per hour) can be calculated using the formula:

   speed = distance travelled ÷ time taken (also written $s = d \div t$)

   a) Calculate Dougie's speed.

   b) Calculate Adam's speed.

2) Customers who buy on hire purchase are charged a deposit and then a fixed monthly payment for a given number of months. The total amount they pay is calculated using this formula:

   cost = deposit + monthly rate x the number of months payments last

   (also written $c = d + mn$)

   Look at these offers.

   a) Use the formula above to calculate the cost of buying this scooter.

   > **Buy now pay later**
   >
   > Drive this scooter away for just £150 deposit and 22 monthly payments of £149

   b) Use the formula above to calculate the cost of booking this holiday.

   > Book next year's luxury holiday cruise for as little as £150 deposit and 10 monthly payments of £199

   c) Use the formula above to calculate the cost of buying this computer.

   > **Pay a deposit of £100 and take this computer home today. Payments of £99 for 15 months**

3) Find the area of this trapezium, using the formula

   area = $\frac{1}{2}$ x parallel sides added together x height

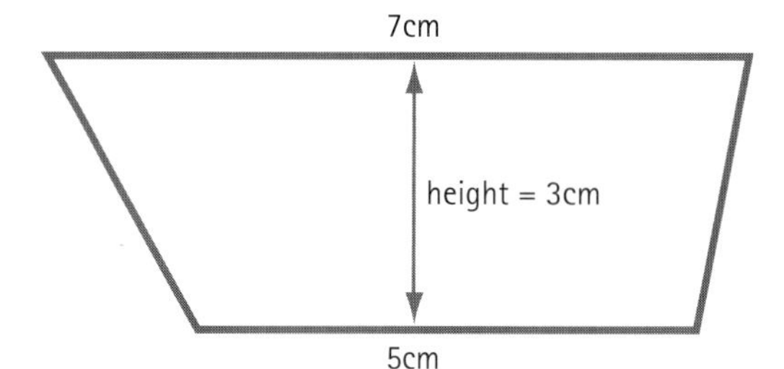

Number patterns

In this unit you will revise:

- understanding number patterns
- number patterns represented by diagrams
- finding the *n*th term

Understanding number patterns

A number pattern is a series or chain of numbers that follow a rule.

- A number pattern may have a rule that **adds** or **subtracts** the same number each time.

  The series 16, 12 , 8,... follows the rule 'subtract 4 each time'. The next number in the series would be 8 – 4 = 4

- A number pattern may have a rule to **multiply** or **divide** by the same number each time.

  The series 16, 8, 4,... follows the rule 'divide by 2 each time'. The next number in the series would be 4 ÷ 2 = 2

- Write the rules for these number chains and fill in the gaps.

a) 3 → 6 → 9 → 12 → ...   b) 1 → 2 → 4 → 8 → ...

Number patterns represented by diagrams

A number pattern can be represented by diagrams. You will need to count shapes to find the rule.

Pattern 1

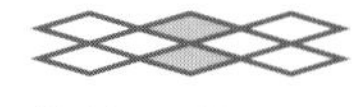
Pattern 2

Pattern 3

See how the pattern is growing.

Pattern 1 has 1 grey tile and 2 white tiles, 3 tiles altogether.
Pattern 2 has 2 grey tiles and 4 white tiles, 6 tiles altogether.
Pattern 3 has 3 grey tiles and 6 white tiles, 9 tiles altogether.

The series of numbers that shows how many grey tiles are needed is:
1, 2, 3...   The rule for the grey tiles is 'add one'.

For pattern 4, you need 4 grey tiles. For pattern 5, you need 5 grey tiles.

The series of numbers that shows how many white tiles are needed is:
2, 4, 6...   The rule for the white tiles is 'add two'.

For pattern 4, you need 8 white tiles. For pattern 5, you need 10 white tiles.

A pattern that you should know is the series of square numbers.

1	4	9	16

Notice 1 x 1 = 1 2 x 2 = 4 3 x 3 = 9 4 x 4 = 16

These numbers are sometimes called
1 squared or 1^2 2 squared or 2^2 3 squared or 3^2 4 squared or 4^2

The series grows like this.

1	4	9	16	25	36	49	64	81	100 ...
1 x 1	2 x 2	3 x 3	4 x 4	5 x 5	6 x 6	7 x 7	8 x 8	9 x 9	10 x 10

Finding the *n*th term

Sometimes, rather than finding the next number or term in a series or chain of numbers, you need to find the 52nd term, or 103rd term, say. Writing out 52 or 103 numbers in a series takes a long time, so try to find a general rule or a general term. This general term is called the *n*th term.

REMEMBER The general term for all times tables can be written in terms of *n*.

Example 1: 3 6 9 12... The rule is 'add 3'.

+ 3 + 3 + 3 This is the 3 times table:

the 1st term is 3 = 3 x 1;
the 2nd term is 6 = 3 + 3 = 3 x 2;
the 3rd term is 9 = 3 + 3 + 3 = 3 x 3;
the 4th term is 12 = 3 + 3 + 3 + 3 = 3 x 4;
so the *n*th term = 3 x n and $3n$ is the general term.
You can find any term you want by substituting for n in the general term.

If you want to find the 5th term $n = 5$ $3n = 3 \times 5 = 15$
If you want to find the 52nd term $n = 52$ $3n = 3 \times 52 = 156$

Example 1

The general term for the 2 times table is $2n$. To find the 4th term of the 2 times table put $n = 4$.

$2n = 2 \times 4 = 8$

Example 2

The general term for the 7 times table is $7n$. To find the 4th term of the 7 times table put $n = 7$.

$7n = 7 \times 4 = 28$

Example 3

6 → 11 → 16 → 21 (+ 5, + 5, + 5) the rule is 'add 5'.

You can see that this is not the 5 times table, but it is close! The general term for the 5 times table is $5n$.

5 → 10 → 15 → 20 ... $5n$ (+ 5, + 5, + 5)

6 11 16 21 ... $5n + 1$

The general rule or nth term for this series of numbers is $5n + 1$
If we want to find the 5th term: $n = 5$ so $5n + 1 = (5 \times 5) + 1 = 26$

Example 4

2 → 6 → 10 → 14 ... (+ 4, + 4, + 4) The rule is 'add 4'.

You can see that this is not the 4 times table, but it is close.

4 → 8 → 12 → 16 ... $4n$ (+ 4, + 4, + 4)

2 6 10 14 ... $4n - 2$

The general rule or nth term for this series of numbers is $4n - 2$

If you want to find the 20th term: $n = 20$ so $4n - 2 = 4 \times 20 - 2 = 78$

Practice questions

1) Copy the number series and number chain. Fill in the gaps and write down the rule.

   a) 66, ..., 44, 33, 22...

   b) 100 000 → 10 000 → ... → 100 → ...

2) Jeannie has white tiles and Gary has grey tiles. Together they make this series of patterns.

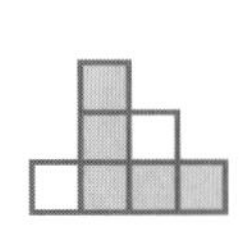

pattern 1

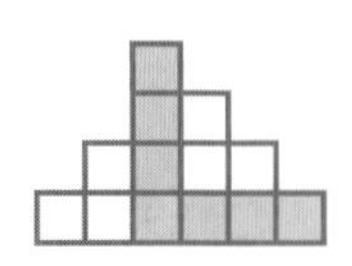

pattern 2

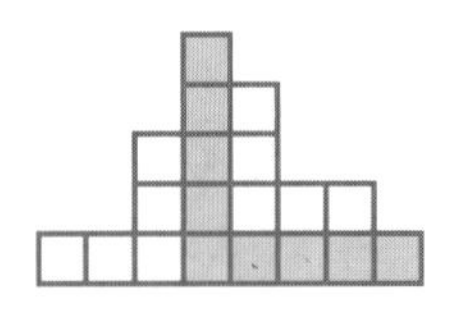

pattern 3

   a) How many white tiles does Jeannie add each time?

   b) How many grey tiles does Gary add each time?

   c) Jeannie and Gary make pattern 4.

   How many white tiles will Jeannie need?
   How many grey tiles will Gary need?

   d) Draw pattern 4 to help Jeannie and Gary with the layout of their tiles.

   e) Pattern 1 uses 7 tiles, pattern 2 uses 13 tiles, pattern 3 uses 19 tiles. 7, 13, 19...

   Find the nth term of this sequence.

   f) Use your nth term to work out how many tiles Jeannie and Gary will need for their 100th pattern.

Equations

In this unit you will revise:

- understanding inverse operations
- solving linear equations
- using equations to solve word problems

Understanding inverse operations

Finding the inverse of a mathematical operation means doing the opposite.

So if the mathematical operation is 'add', the inverse operation is 'subtract'
if the mathematical operation is 'subtract', the inverse operation is 'add'
if the mathematical operation is 'multiply', the inverse operation is 'divide'
if the mathematical operation is 'divide', the inverse operation is 'multiply'

Example 1

Start with the number 5, use the rule + 4 on it to get 9 $5 \rightarrow +4 \rightarrow 9$

Now take the number 9, use the inverse rule − 4 on it to get back to the given number: $9 \rightarrow -4 \rightarrow 5$

Example 2

Start with the number 6, use the rule x 2 + 3 on it to get 15

$6 \rightarrow \times 2 \rightarrow +3 \rightarrow 15$

Now take the number 15, use the inverse of the rule − 3 ÷ 2 on it to get back to the given number: $15 \rightarrow -3 \div 2 \rightarrow 6$

- Write the inverse of these rules.: a) ÷ 8 b) + 7 c) x 4 − 1

REMEMBER
A formula is a rule that is used on a given number to give a new number: given number → rule → new number.
If you use the inverse of the rule on the new number you get back to the given number: new number → inverse of rule → given number

REMEMBER
Turn to page 27 for help with understanding formulae.

Using inverse operations to solve linear equations

Linear equations have one unknown number written as a letter. This is one way to solve them:

- read the equation and decide what rule is being worked on the unknown
- apply the inverse rule to find the value of the unknown
- check your answer by putting the value for the unknown into the equation.

REMEMBER There are several ways of solving linear equations. You may have learnt a different method at school. Perhaps you use the balance method shown on the KS3 Bitesize Maths TV programme. If you are confident with the way you know, use your method, but check your final answer with ours.

Example 1 Find the value of x if $x + 4 = 9$

- Given the unknown number x, using the rule + 4 on it gives 9
  $x \rightarrow + 4 \rightarrow 9$
- Apply the inverse rule to find x: $9 \rightarrow - 4 \rightarrow 5$. So x is 5.
- Check the answer: if $x + 4 = 9$ and x is 5, then check $5 + 4 = 9$ ✓

Example 2 Find the value of x if $5x - 1 = 14$

- Given the unknown number x, using the rule x 5 – 1 on it gives 14
  $x \rightarrow \times 5 - 1 \rightarrow 14$
- Apply the inverse rule to find x: $14 \rightarrow + 1 \div 5 \rightarrow 3$. So x is 3.
- Check the answer: if $5x - 1 = 14$ and x is 3, then check $15 - 1 = 14$ ✓

- Solve these linear equations.

a) $8y = 48$ b) $9r - 12 = 6$ c) $\frac{a + 23}{4} = 7$

Using equations to solve word problems

Equations can be used to solve many everyday problems.

Example 1

Julie ordered an enormous cake for her brother's 21st birthday. She bought 2 boxes of birthday candles (with the same number of candles in each box) and 5 single birthday candles, making a total of 21 candles.

a) Represent this situation as a linear equation.

b) Solve the linear equation to find how many birthday candles in a box.

The number of birthday candles in a box is the unknown number. Call the number of candles in a box c. Now read the problem again.

Julie buys 2 boxes of candles and 5 single candles, making 21 candles in all.

a) $2 \times c + 5 = 21$ So the linear equation is $2c + 5 = 21$
b) Given the unknown number c, using the rule x 2 + 5 on it, gives 21
$c \rightarrow \times 2 + 5 \rightarrow 21$

Apply the inverse rule to find c: $21 \rightarrow - 5 \div 2 \rightarrow 8$ so $c = 8$

Check the answer: if $2c + 5 = 21$ and c is 8, then check $2 \times 8 + 5 = 21$ ✓

Therefore, there are 8 birthday candles in each box.

Example 2

Julie's father will not tell anyone how old he is. However, he will say that if you take his age and half it, then subtract 3, you get to his son's age of 21.

a) Represent this situation as a linear equation.

b) Solve the linear equation to find the age of Julie's father.

Julie's father's age is the unknown number. Let's call his age x years.

Now read the problem again.

Take Julie's father's age, half it, then subtract 3 to get his son's age of 21.

a) $x \div 2 - 3 = 21$ So the linear equation is $\frac{x}{2} - 3 = 21$

b) Given the unknown number x, using the rule $\div 2 - 3$ on it gives 21

$x \rightarrow \div 2 - 3 \rightarrow 21$

Apply the inverse rule to find x: $21 \rightarrow + 3 \times 2 \rightarrow 48$ so $x = 48$

Check the answer: if $\frac{x}{2} - 3 = 21$ and x is 48, then check $\frac{48}{2} - 3 = 21$

Therefore, Julie's father is 48 years old.

Practice questions

1) Here are five mathematical operations cards.

x 25	- 5	- 1 ÷ 25	÷25 - 1	÷ 5

a) Which card is the inverse rule for + 5?

b) Which card is the inverse rule for x 25 + 1

c) Which card is the inverse rule for x 5?

d) Which card is the inverse rule for ÷ 25?

e) Which card is the inverse rule for + 1 x 25

2) Solve these equations.

a) $n + 3 = 8$

b) $4x = 24$

c) $4p + 3 = 11$

3) I think of a number, n. If I take the number, n, and multiply by 3, then add 1, I get the answer 46.

a) represent this situation as a linear equation using n.

b) solve the linear equation to find the number, n, I am thinking of.

Coordinates, mappings and line graphs

In this unit you will revise:

- reading and plotting coordinates
- mappings
- finding the equations of straight lines

REMEMBER An x is a cross, so the x-axis goes across →.

Reading and plotting coordinates

The horizontal axis is the x-axis. The vertical axis is the y-axis.

REMEMBER Coordinates are always written as (x-coordinate, y-coordinate).

When reading or plotting coordinates:

- start at (0,0), the origin
- move across right or left: right → for positive x-numbers; left ← for negative x-numbers
- move up or down: up ↑ for positive y-numbers; down ↓ for negative y-numbers

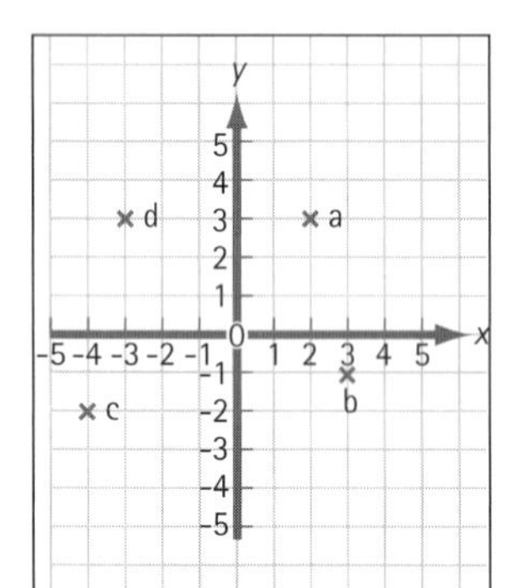

Look carefully at the points marked with crosses.

For point a, start at the origin, move across right → 2 and up ↑ 3 so $x = 2$ and $y = 3$ and the coordinates are (2, 3).

For point b, start at the origin, move across right → 3 and down ↓ 1 so $x = 3$ and $y = -1$ and the coordinates are (3, −1).

For point c, start at the origin, move across left ← 4 and down ↓ 2 so $x = -4$ and $y = -2$ and the coordinates are (−4, −2).

For point d, start at the origin, move across left ← 3 and up ↑ 3 so $x = -3$ and $y = 3$ and the coordinates are (−3, 3)

- Copy the axes above and plot and label the following points.

e(−2, −1) f(3, 2) g(−5, 1) h(−4, −1) i(1, −3)

Mappings

A mapping is a rule that works on a number to give a new number. Mappings may be represented as diagrams.

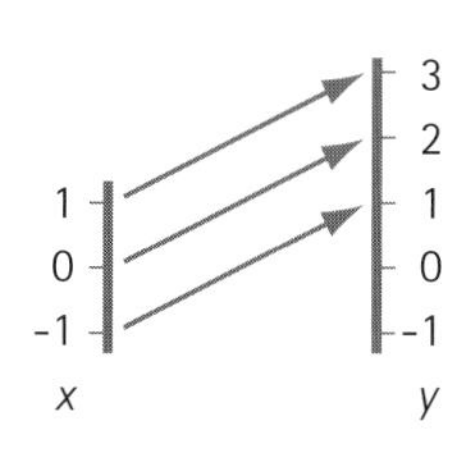

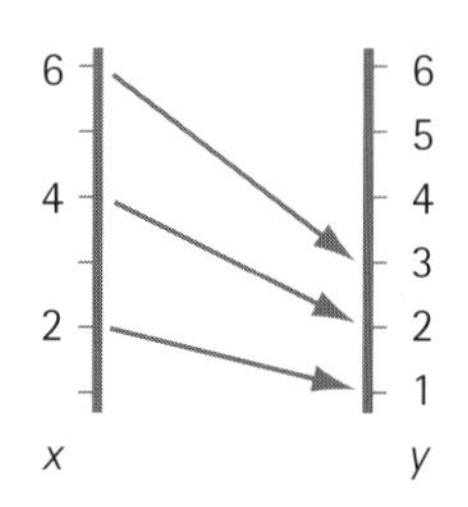

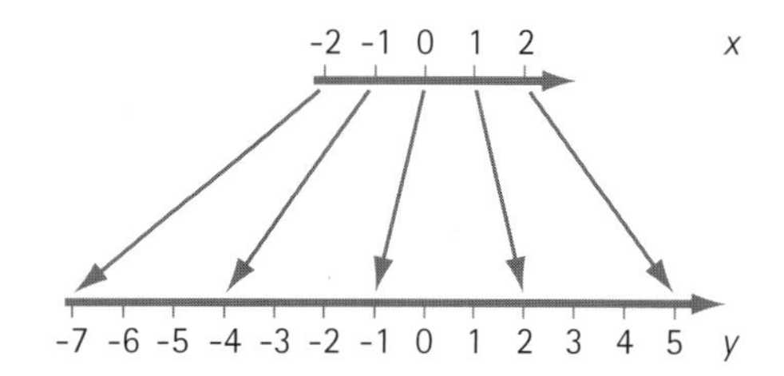

the rule is + 2 the rule is ÷ 2 the rule is x 3 – 1

$x \rightarrow x + 2$ $x \rightarrow \frac{x}{2}$ $x \rightarrow 3x - 1$

so $y = x + 2$ so $y = \frac{x}{2}$ so $y = 3x - 1$

Note that these mappings also give coordinates that can be plotted on a grid.

The mapping for $y = x + 2$ gives coordinates (–1, 1), (0, 2) and (1, 3).

The mapping for $y = \frac{x}{2}$ gives coordinates (2, 1), (4, 2) and (6, 3).

The mapping for $y = 3x - 1$ gives coordinates (–2, –7), (–1, 4), (0, –1), (1, 2), (2, 5).

• Copy and complete the mapping diagram for $x = 2$, $x = 1$ and $x = 0$ and plot the points on the axes. Join the points to make a graph.

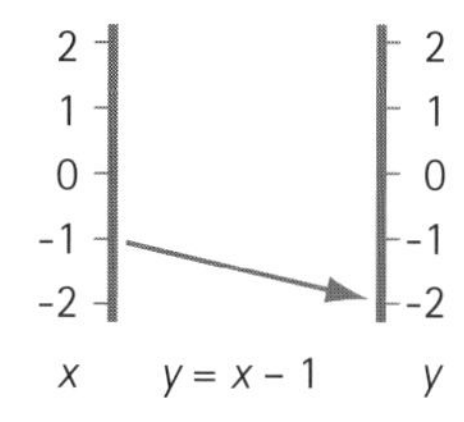

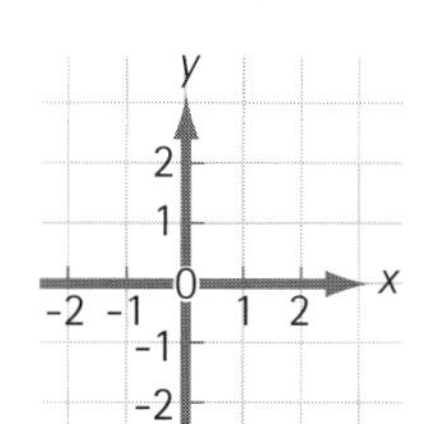

Finding the equations of straight lines

Look at these mapping diagrams.

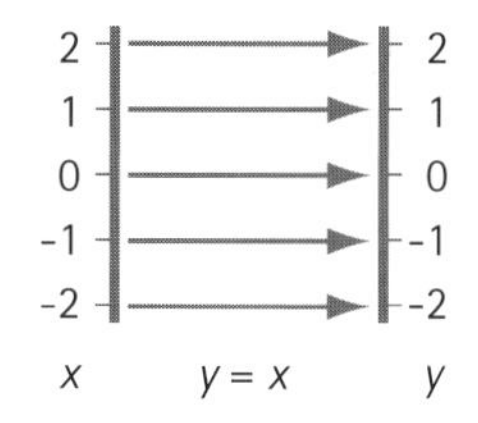

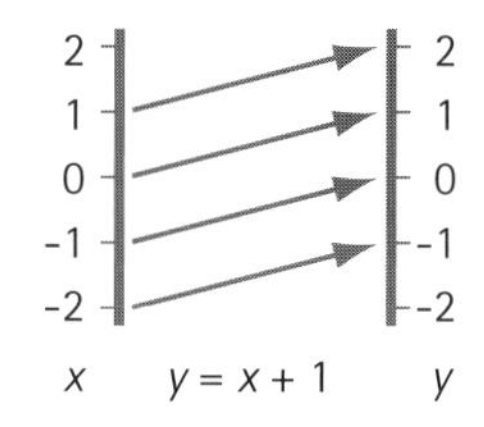

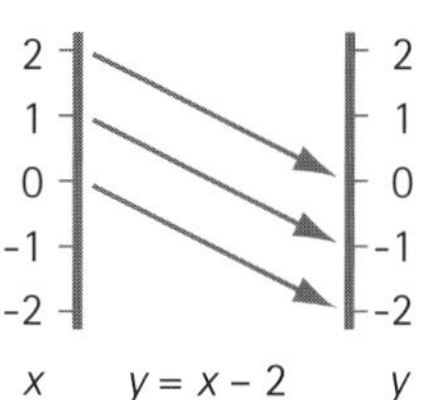

The mappings of these equations give these straight lines.

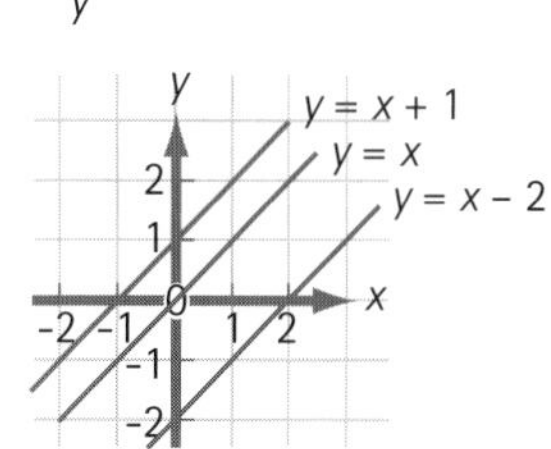

REMEMBER Turn to page 47, to read about parallel lines.

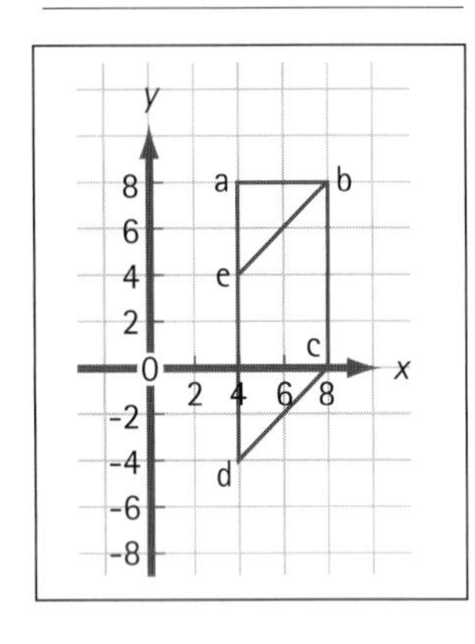

Notice $y = x$ crosses the y-axis at $y = 0$.

Also, notice the lines are parallel.

$y = x + 1$ crosses the y-axis at $y = 1$; $y = x - 2$ crosses the y-axis at $y = -2$.

Where do you think $y = x + 3$, $y = x + 5$ and $y = x - 4$ will cross the y-axis?

They cross at $y = 3$, $y = 5$ and $y = -4$.

Example

Look at this diagram of a right-angled triangle joined to a parallelogram.

The line thorough points a and b has the equation $y = 8$, because the y-coordinate of every point on that line is 8.

The line through points a, e and d has the equation $x = 4$, because the x-coordinate of every point on that line is 4.

The line through points b and e has the equation $y = x$.

- Write the equation for the line through points:

a) b and c b) c and d.

Practice question

1) Look at the pattern of triangular tiles.

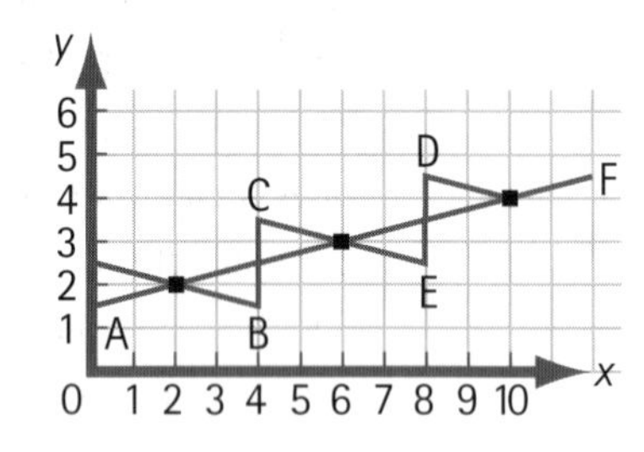

The point at which the triangles touch has been marked with a square.

The third square has coordinates (10, 4).

a) What are the coordinates of the first and second squares?

b) What would be the coordinates of the fourth square?

Angles

In this unit you will revise:

- understanding angles
- measuring angles accurately
- drawing angles accurately
- calculating angles

Understanding angles

An angle measures a turn or a rotation. Angles are measured in degrees (°).

angles measuring 0° to 90° are acute	angles measuring 90° to 180° are obtuse	angles measuring 180° to 360° are reflex

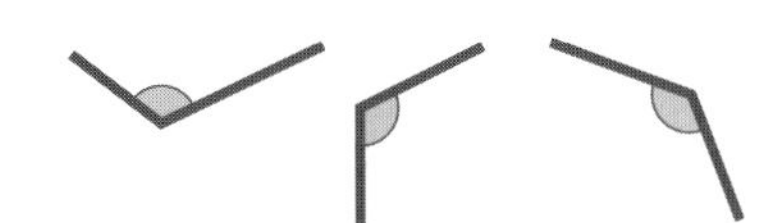

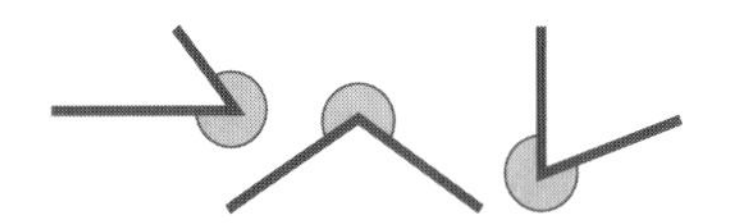

- Are these angles right angled, acute, obtuse or reflex?

a) 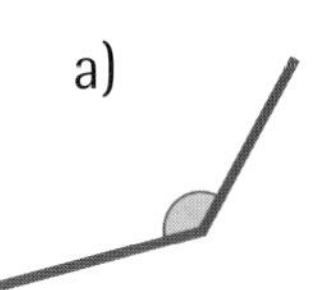

b) 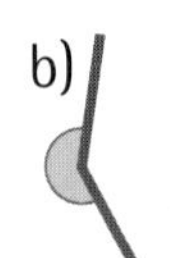

c) 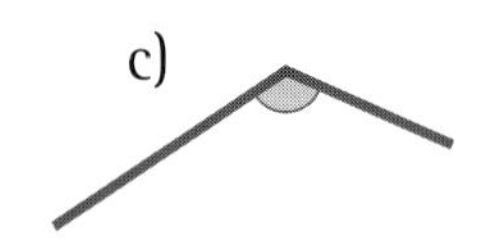

d) 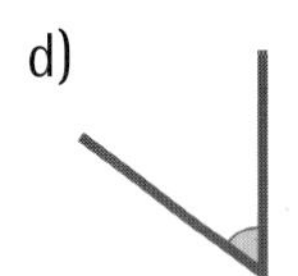

Measuring angles accurately

To measure an angle accurately:

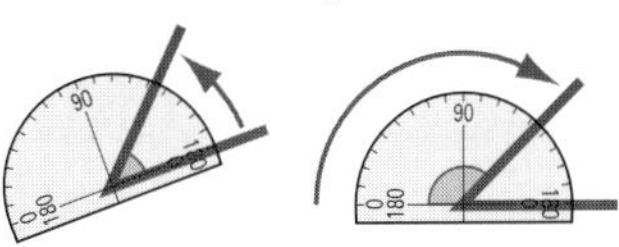

- place your protractor over the angle, with its base line on one of the arms of the angle
- make sure the point of the angle is exactly under the centre of the protractor
- follow the scale from 0 round to where the other arm of the angle meets the scale
- read the number of degrees – you must be accurate to within 1°.

REMEMBER Some protractors have two scales. If yours does, make sure that you use the right scale, by checking that one arm of the angle is under the zero. Check your angle as well, and decide if it is acute, obtuse or reflex. Make sure that your measurement agrees.

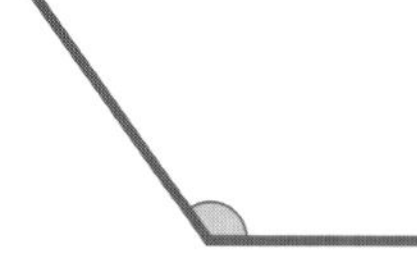

Example

Measure this angle. Notice that it is an obtuse angle so it will be between 90° and 180°.

Place the protractor over the angle like this: or this:

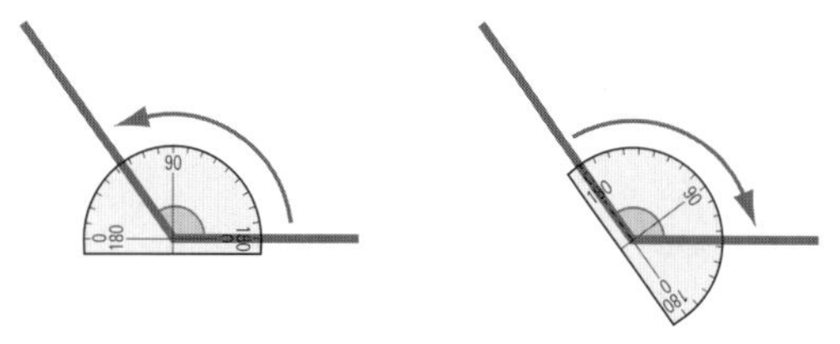

Measure from 0 (in the direction of the arrow).

The angle is 126°.

- Measure these angles.

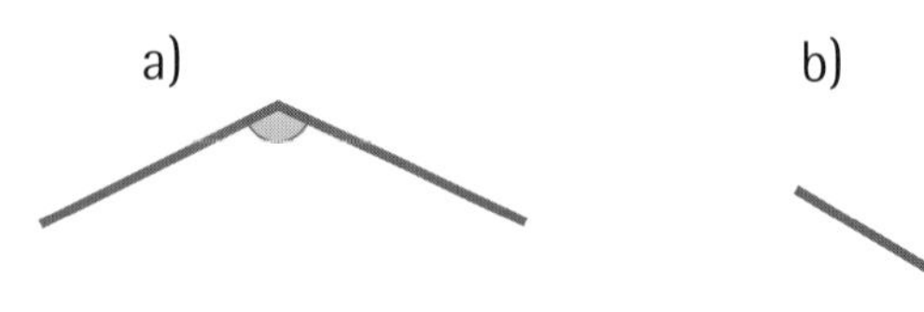

Drawing angles accurately

To draw an angle accurately:

- draw a straight line
- mark a small cross anywhere on your straight line
- place your protractor with its centre over the x on your line and 90° is directly above the x

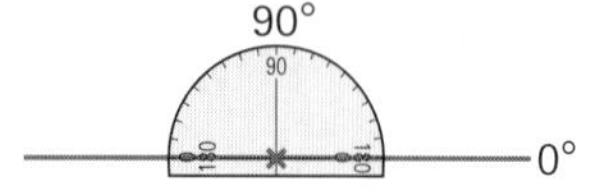

- follow the scale from 0 round to find the angle you need

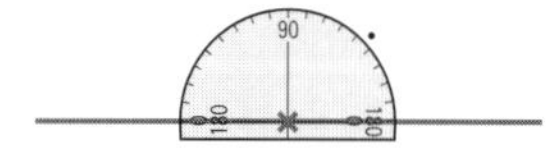

- mark the point with a dot – your angle must be accurate to within 1° of the required angle
- move the protractor away, join the cross and the dot and mark the angle.

- Draw these angles. a) 40° b) 157°

FactZONE

Calculating angles

Factzone

- There are 360° in a complete turn, like the angle at the centre of a circle.

  $a + b = 360°$

- There are 180° in a half turn, the angle makes a straight line.

  $a + b = 180°$

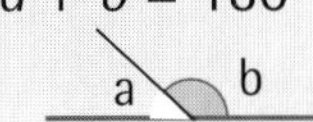

- There are 90° in a quarter turn, the angle is a right angle.

  $a + b = 90°$

Example 1

Find the value of x.

Notice that the two angles together make a complete turn of 360° so $x = 360° - 257° = 103°$

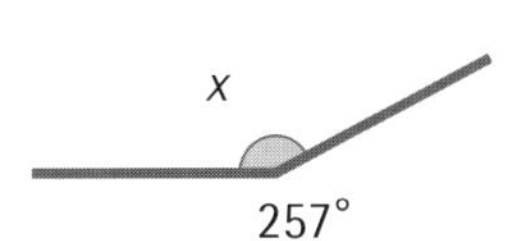

Example 2

Find the value of y.

Notice that the two angles make a half turn of 180°

so $y = 180° - 79° = 101°$

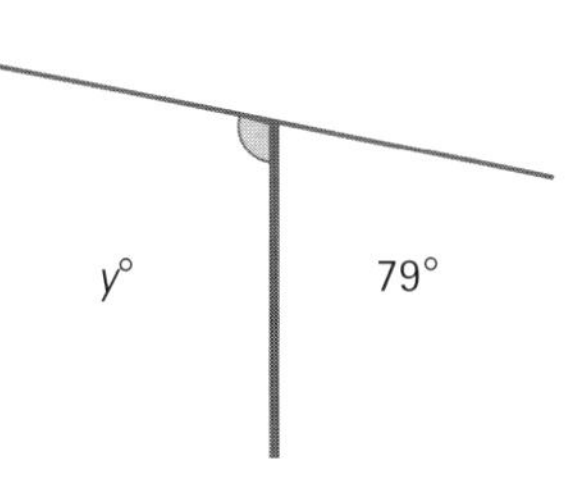

Example 3

Find the value of z.

Notice that the two angles make a quarter turn of 90°

so $z = 90° - 52° = 38°$

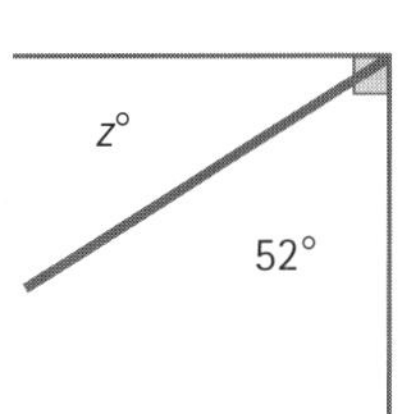

- Calculate the size of these angles. Give reasons for your answers.

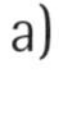

a)

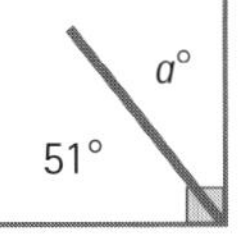

b)

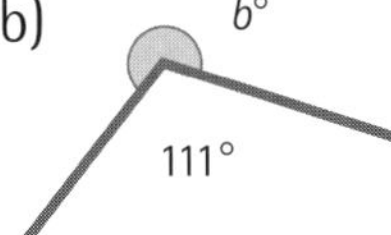

c)

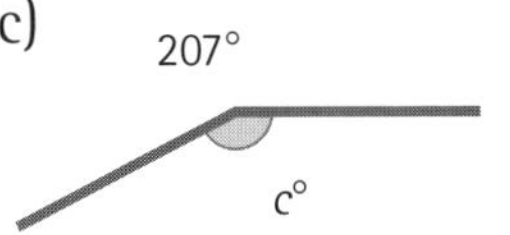

d)

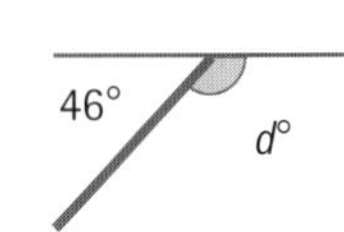

Practice questions

1) Measure these angles.

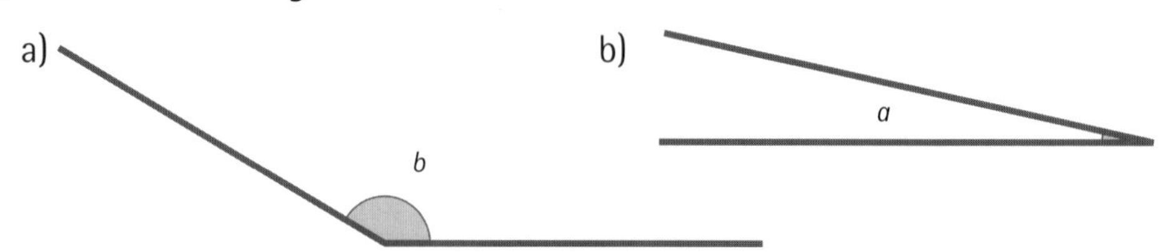

2) Draw angles of a) 63° b) 137° c) 241°.

3) Calculate the missing angles, giving reasons for your answers.

Not drawn to scale

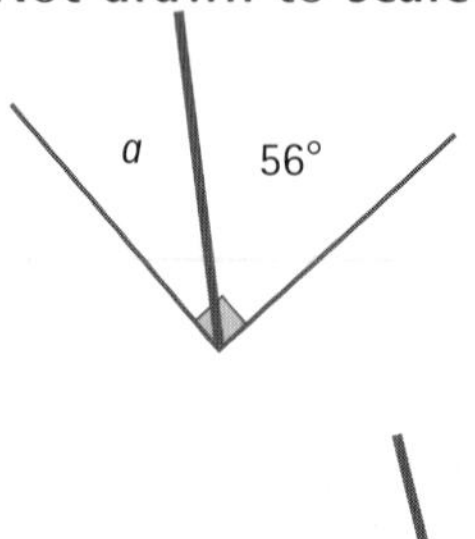

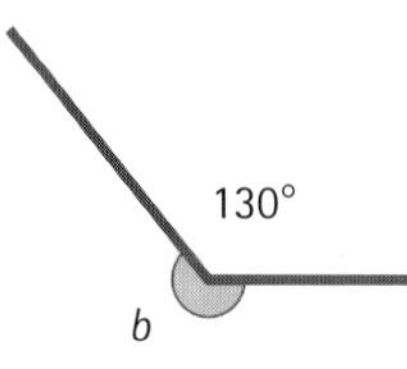

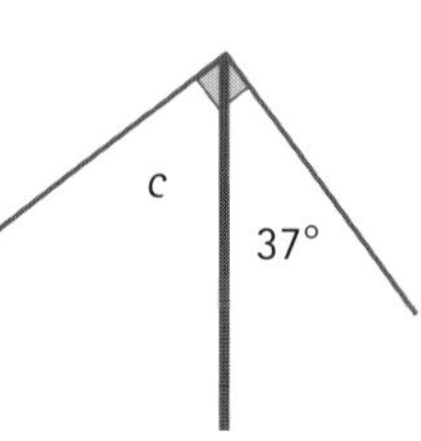

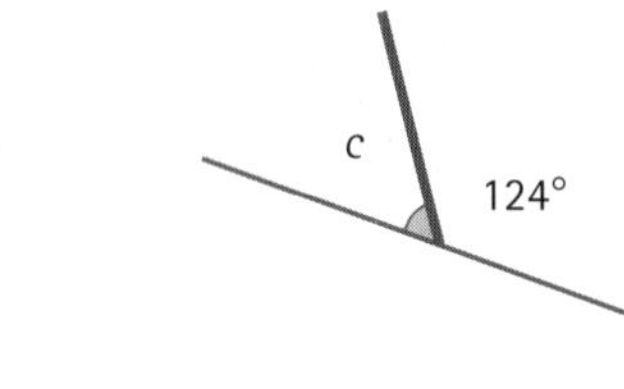

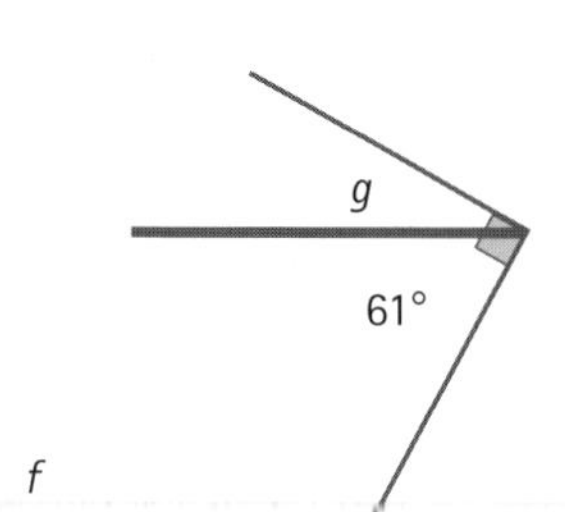

Parallels and angles

In this section you will revise:

- understanding parallel lines
- intersecting lines

Understanding parallel lines

Lines are parallel if they are always the same perpendicular distance apart.

This means they never meet.

Example 1

These three lines are parallel.

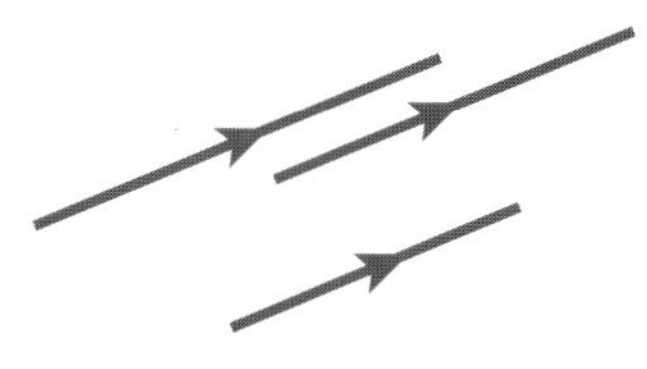

Example 2

These two lines are parallel. The small arrows on the lines show they are parallel.

Look at this parallelogram.

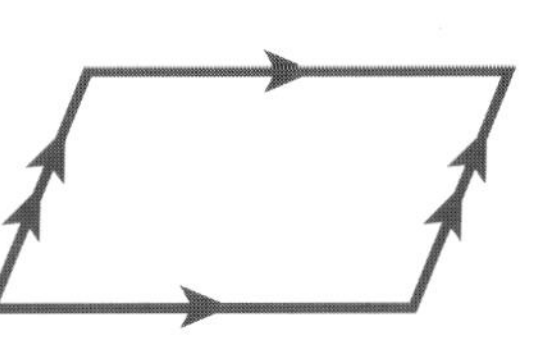

The two sides with one arrow are parallel.

The two sides with two arrows are parallel.

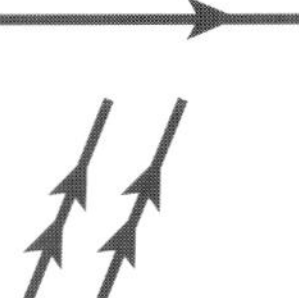

Intersecting lines

A transversal is a straight line that intersects (cuts) parallel lines.

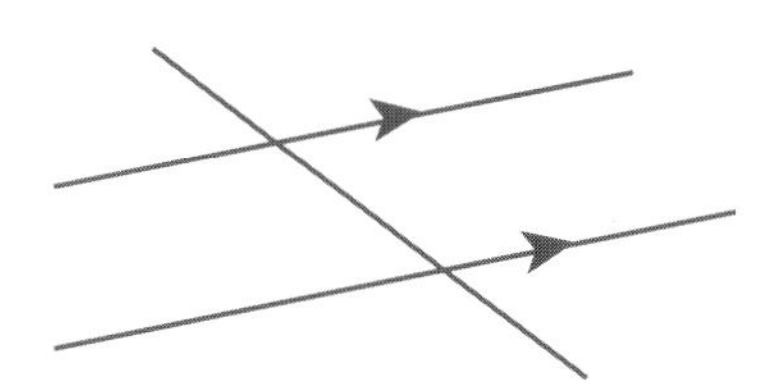

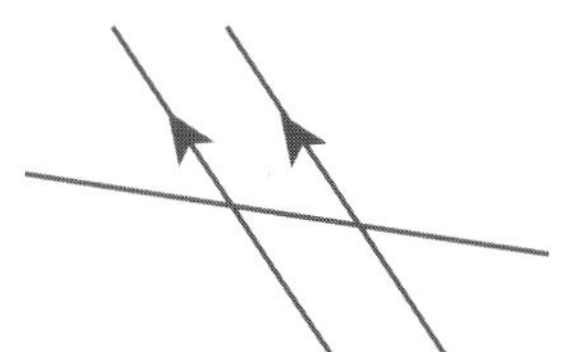

Parallels and angles

FactZONE

Corresponding angles make an F-shape

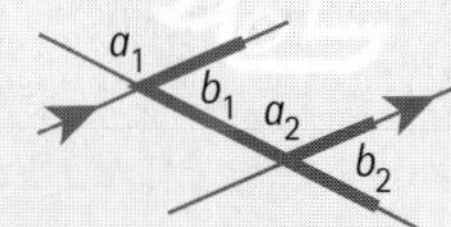

corresponding angles are equal
$a_1 = a_2$ and $b_1 = b_2$

Alternate angles make a Z-shape

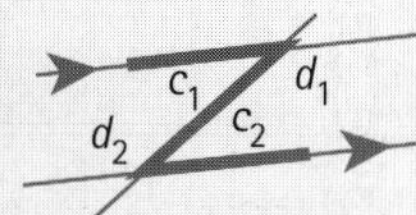

alternate angles are equal
$c_1 = c_2$ and $d_1 = d_2$

Interior angles make a U-shape

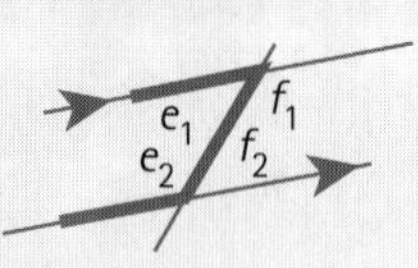

interior angles add up to 180°
$e_1 + e_2 = 180°$ and $f_1 + f_2 = 180°$

When two straight lines intersect they create the following angles.

vertically opposite angles make an X-shape

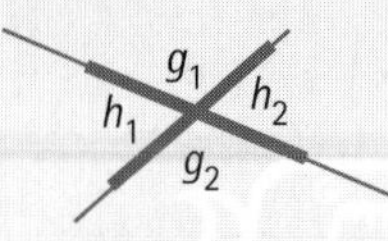

vertically opposite angles are equal
$g_1 = g_2$ and $h_1 = h_2$

Example 1

Find the value of p.

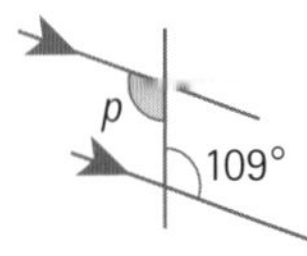

The two angles are alternate
so $p = 109°$ (alternate angles)

Example 2

Find the value of q.

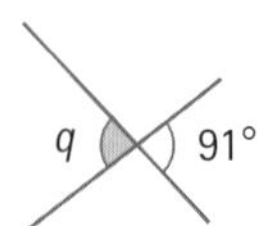

The two angles are vertically opposite
so $q = 91°$ (vertically opposite angles)

Example 3

Find the value of r.

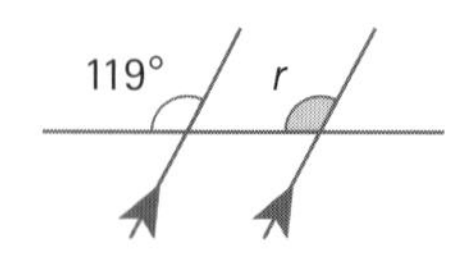

The two angles are corresponding
so $r = 119°$ (corresponding angles)

Example 4

Find the value of s.

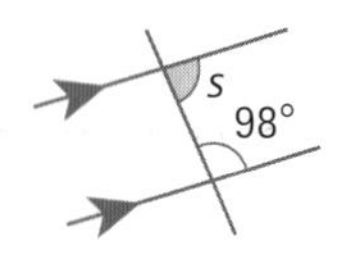

The two angles are interior
$s = 180° - 98° = 82°$ (interior angles)

- Find the missing angles (giving reasons for your answers).

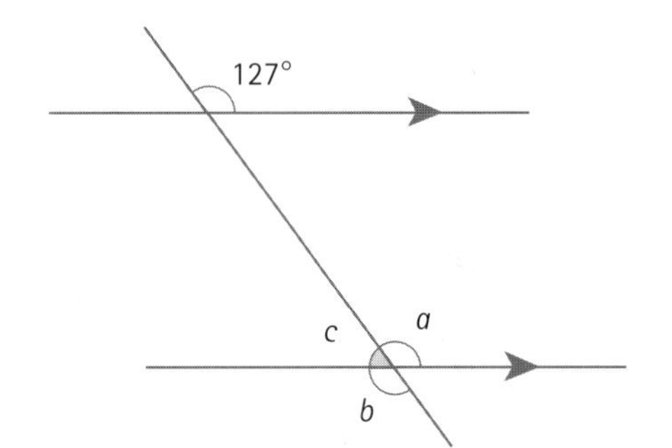

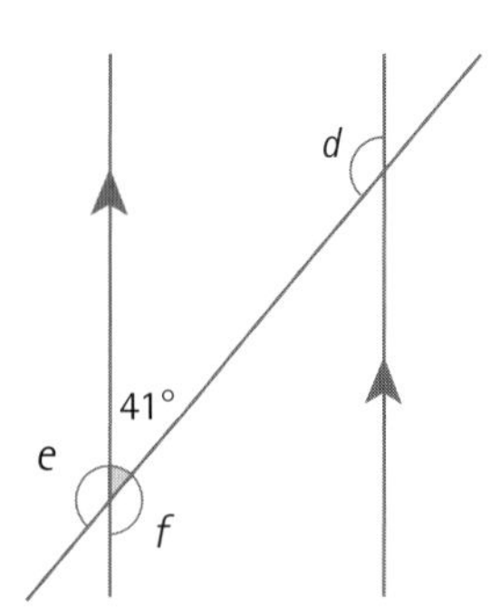

Parallels and angles

Look at these diagrams.

Find the missing angle in each diagram, giving reasons for your answers.

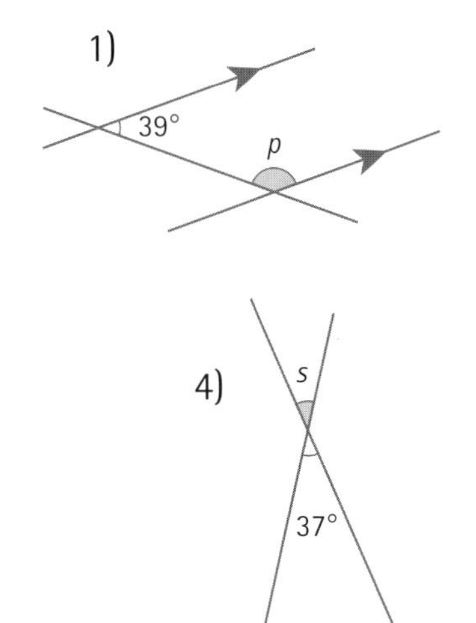

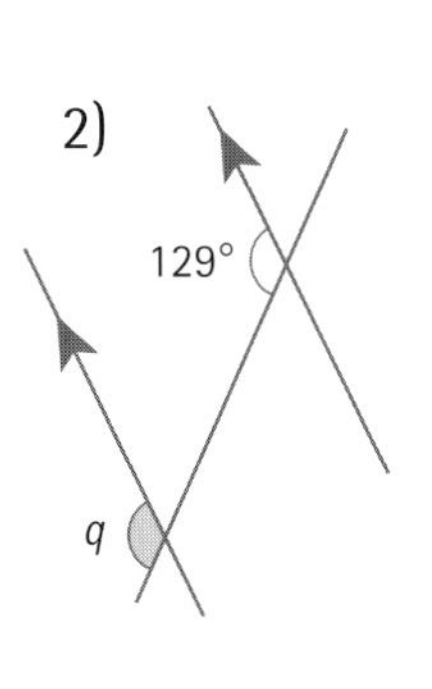

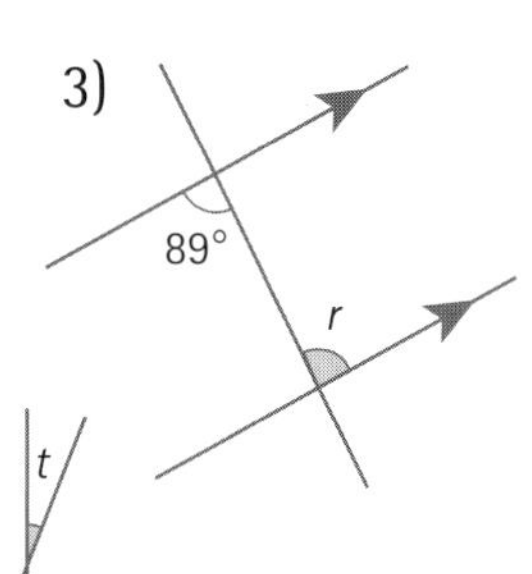

5)

t

22°

2D shapes and 3D solids

In this unit you will revise:

- two-dimensional shapes
- three-dimensional solids
- two-dimensional views of three-dimensional solids
- nets

Two-dimensional shapes

Two-dimensional shapes are flat, they have length and width but no depth.

Example

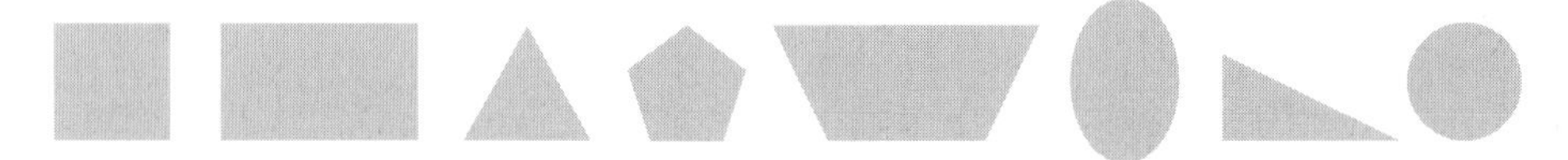

Three-dimensional solids

Three-dimensional solids have length, width and depth. They appear to rise out of the paper they are drawn on.

Example

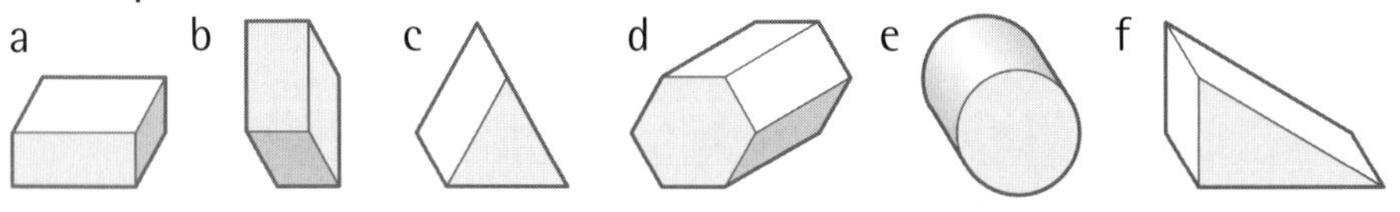

Drawing three-dimensional solids is easier if you use isometric paper.

Here are the three-dimensional solids a and f drawn on isometric paper

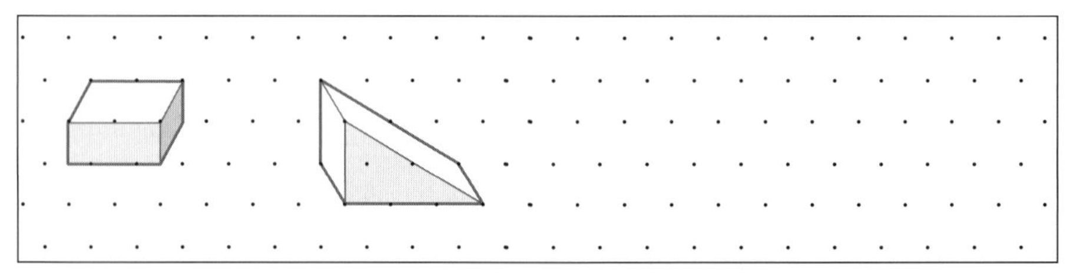

- On a separate piece of isometric paper draw solids c and d.

Two-dimensional views of three-dimensional solids

Three-dimensional solids have two-dimensional faces.

Example 1

This cuboid has four rectangular faces and two square faces.

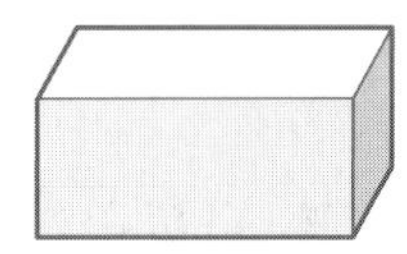

looking down on to the top (top elevation) you see the rectangle

looking on to the side (side elevation) you see the square

looking on to the front (front elevation) you see the rectangle

Example 2

This cylinder has two circular faces and one curved surface

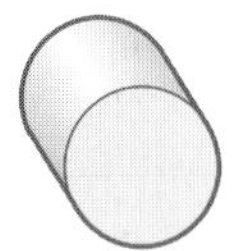

The top elevation is a rectangle

The front elevation is a circle

The side elevation is a rectangle

- Write what two-dimensional shapes you see from the top, front and side elevations of this solid.

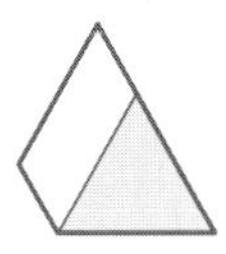

Nets

A net is a two-dimensional pattern for a three-dimensional solid.

Example 1

This cuboid

has this net.

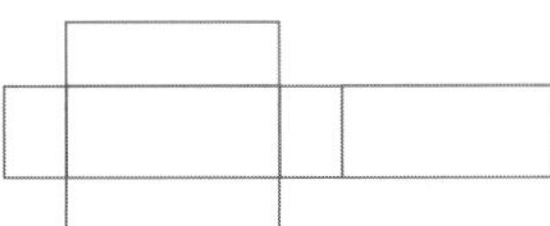

Example 2

This triangular prism

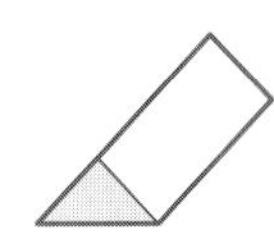

has this net:

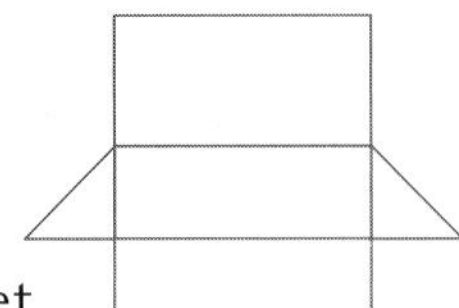

Solids can have more than one net.

The cuboid in example 1 could also have this net.

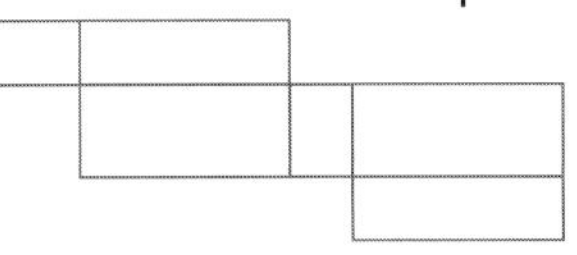

The triangular prism in example 2 could also have this net.

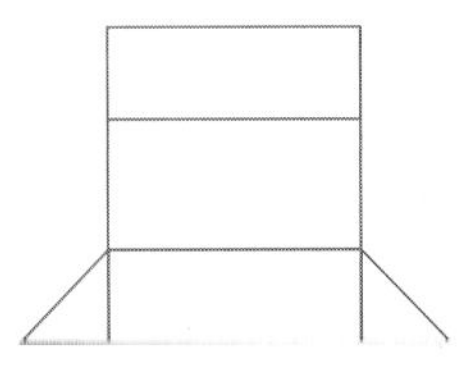

The net for this cylinder looks like this:

Notice, the length of the rectangle should be the same as the circumference of each circle.

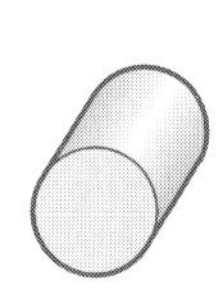

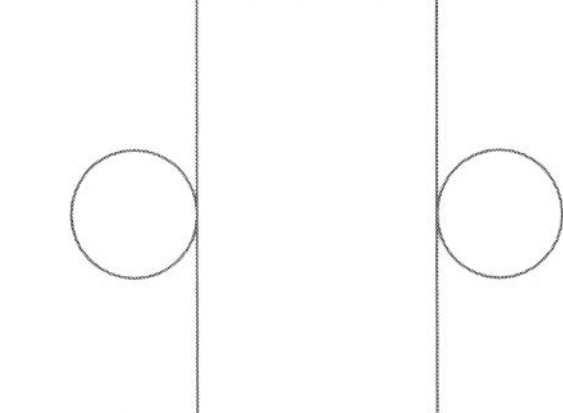

Practice questions

Look at this three-dimensional solid.

1) Draw a net of the solid.

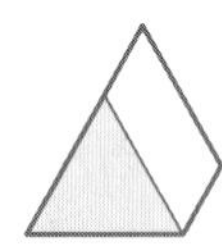

2) Draw the solid on the isometric paper.

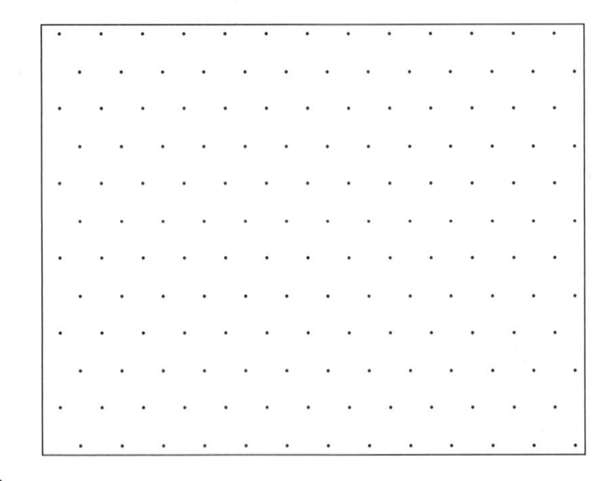

3) The solid is used to make this model.

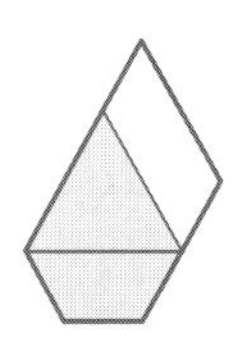

Draw

a) the side elevation

b) the front elevation

c) the top elevation.

Polygons

In this section you will revise:

- understanding polygons
- quadrilaterals
- angles of a polygon

Understanding polygons

Polygons are closed 2D shapes with straight sides. These are polygons:

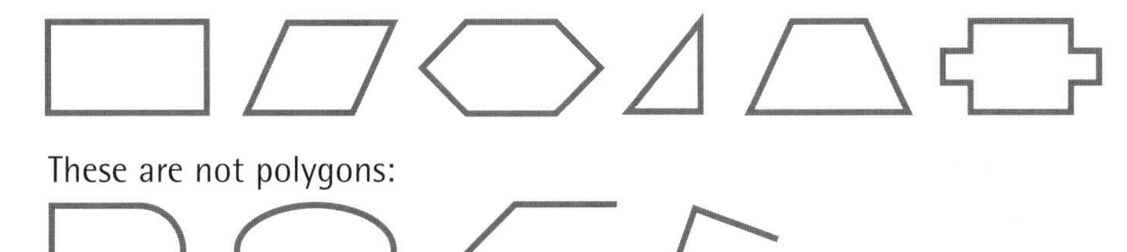

These are not polygons:

because they do not have straight sides and they are not closed.
Regular polygons have all sides equal and all angles equal. These are regular polygons:

square

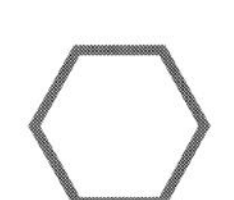

equilateral triangle

regular hexagon

Polygons that are not regular are called irregular polygons. These are irregular polygons:

rectangle

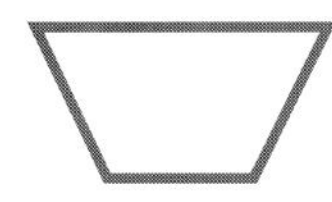

trapezium

Quadrilaterals

Quadrilaterals are polygons with four sides.

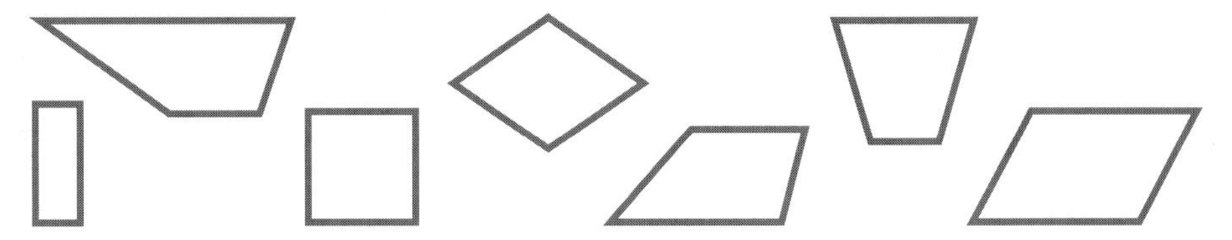

REMEMBER A closed shape is one in which all the sides join up and there are no unmatched ends.

REMEMBER Every quadrilateral has four sides, four vertices (corners) and two diagonals. Special quadrilaterals have special properties. You need to know about their sides, their angles and their diagonals.

FactZONE

■ Square

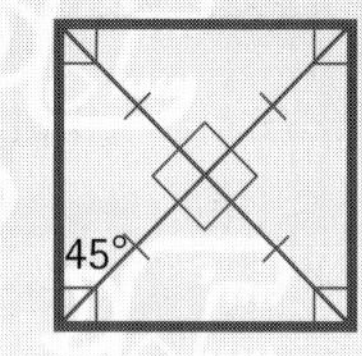

A square is a special rectangle.

All sides are equal and opposite sides are parallel. All angles are 90°.

The diagonals are equal. They bisect (cut) each other at right angles (90°) and bisect the right angles at each vertex (corner) to make angles of 45°.

■ Rhombus

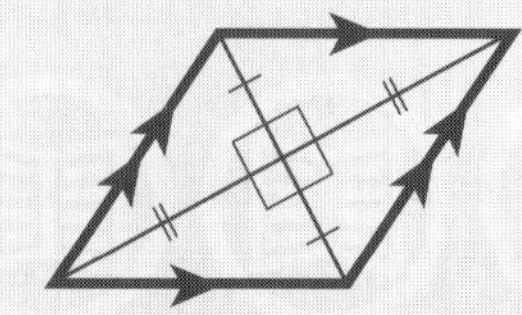

A rhombus is like a squashed square.

All sides are equal and opposite sides are parallel.

Opposite angles are equal.

The diagonals are not equal. They bisect each other at right angles.

■ Rectangle

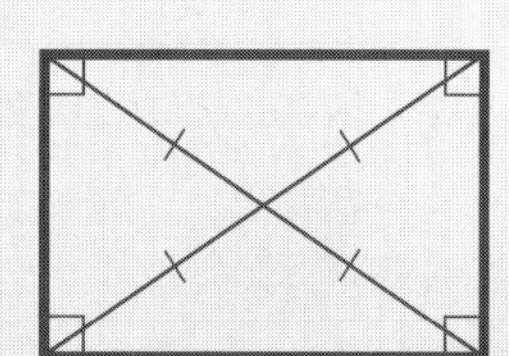

Opposite sides are equal and parallel. All angles are 90°.

The diagonals are equal. They bisect each other.

■ Parallelogram

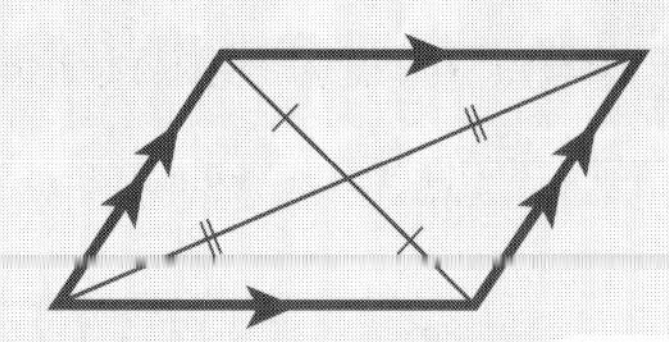

A parallelogram is like a squashed rectangle.

Opposite sides are equal and parallel. Opposite angles are equal.

The diagonals are not equal. They bisect each other.

■ Kite

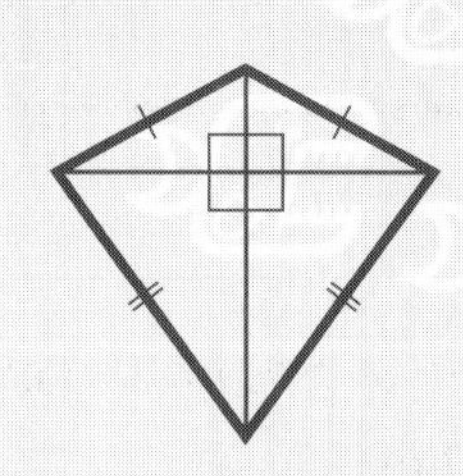

A kite is like a diamond shape, but one point is longer than the other.

Two pairs of adjacent sides (sides next to each other) are equal.

The diagonals are not equal. They bisect each other at right angles.

■ Trapezium

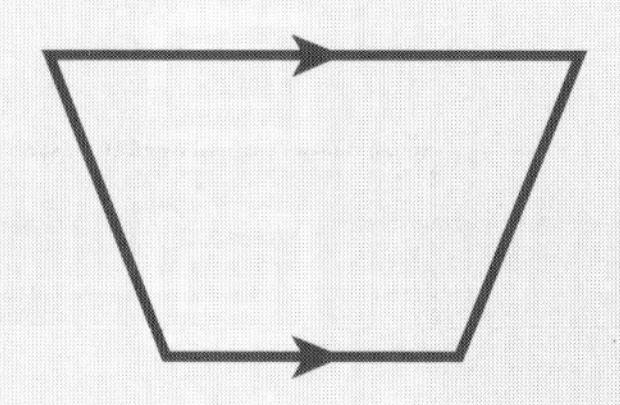

A trapezium is a quadrilateral with one pair of parallel sides.

Notice that some of the sides of the quadrilaterals above have tiny lines or arrows drawn on them. Remember, the lines show equal sides and the arrows show parallel sides.

Angles of a polygon

- Angles on a straight line add up to 180°.

- Angles in a triangle also add up to 180°.

Look at this triangle. Separate out the angles. Rearrange them to make a straight line.

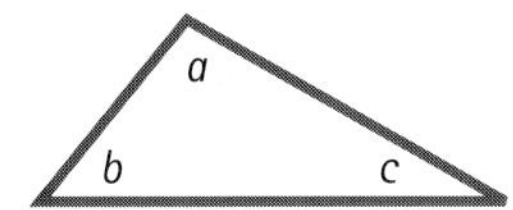

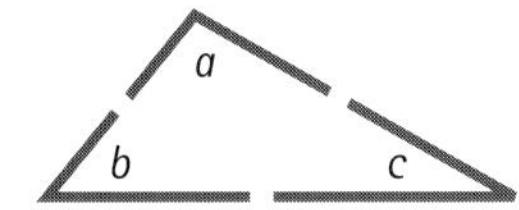

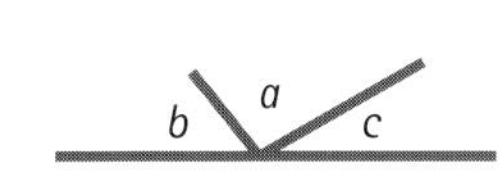

- Angles inside vertices of a polygon are interior angles.

You can find the sum of the interior angles for polygons by looking at triangles.

Example 1

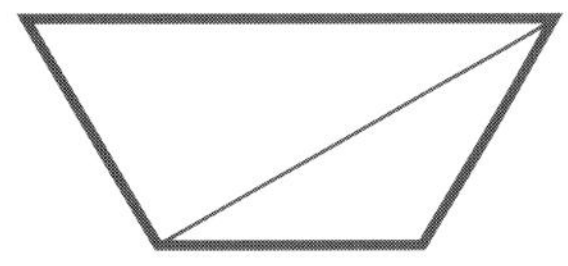

This trapezium splits into two triangles,

so the sum of the interior angles of a trapezium = 2 x 180° = 360°

Example 2

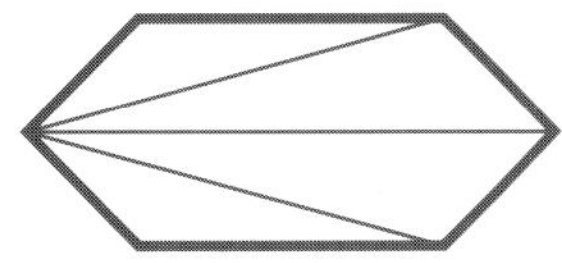

This hexagon splits into four triangles,

so the sum of the interior angles of a hexagon = 4 x 180° = 720°

- Write the sums of the interior angles of these polygons.

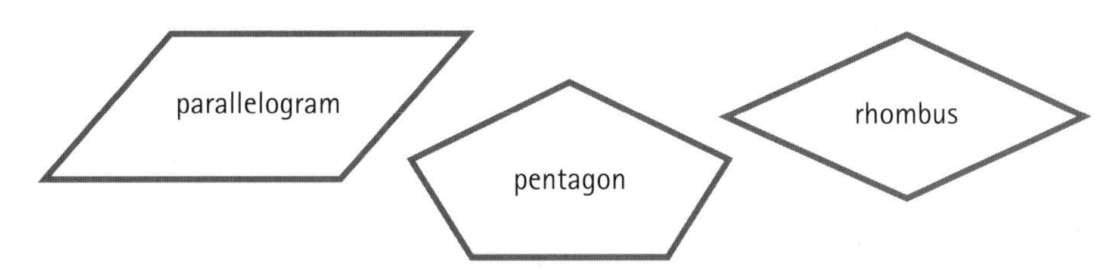

The angles on the outside of the vertices of a polygon are called exterior angles.

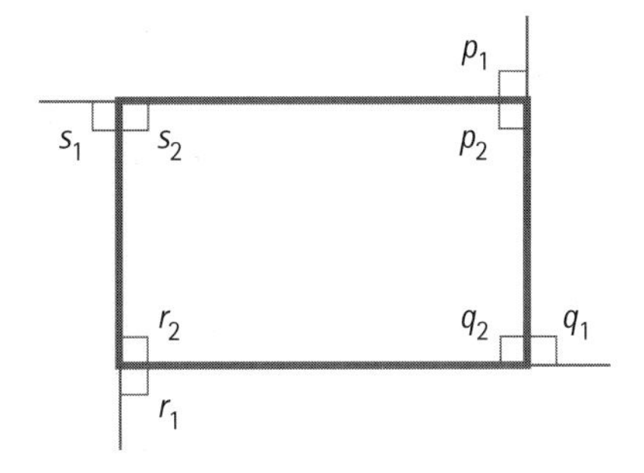

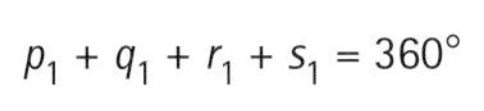

$p_1 + q_1 + r_1 + s_1 = 360°$

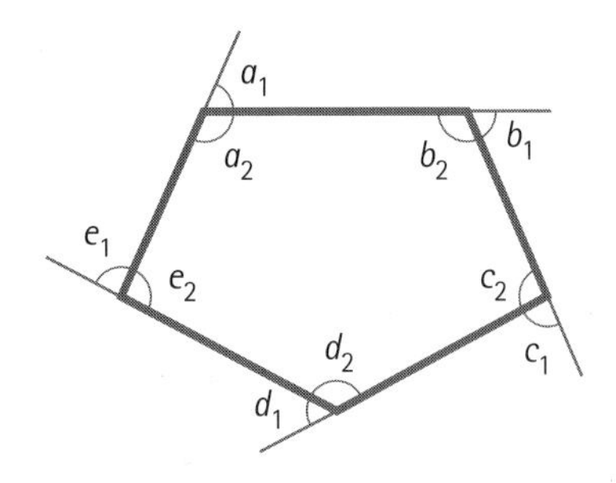

$a_1 + b_1 + c_1 + d_1 + e_1 = 360°$

REMEMBER This seems like a lot to learn but don't panic. You already know a lot. Close the book and write down everything you know about sides, angles and diagonals of: squares, rhombuses, rectangles, parallelograms, kites and trapeziums - there isn't so much left to learn!

Polygons

REMEMBER See calculating angles, page 45, to read about angles on a straight line.

The sum of the exterior angles for any polygon is 360°, because together they make a full turn.

$p_1 + q_1 + r_1 + s_1 = 360°$ $a_1 + b_1 + c_1 + d_1 + e_1 = 360°$

At any corner or vertex of a polygon

interior angle + exterior angle = 180°

so $p_1 + p_2 = 180°$, $q_1 + q_2 = 180°$...

$a_1 + a_2 = 180°$, $b_1 + b_2 = 180°$...

Example 3

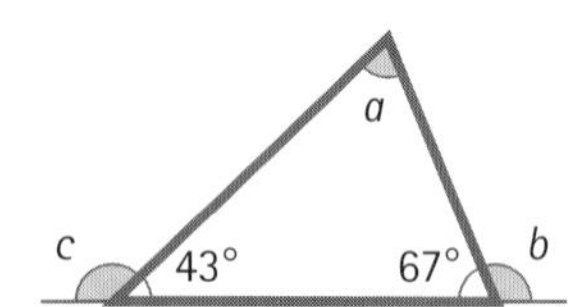

Find the missing angles.

$a = 180° - 67° - 43° = 70°$ (sum of interior angles of triangle)

$b = 180° - 67° = 113°$

$c = 180° - 43° = 137°$

- Calculate the missing angles in this trapezium.

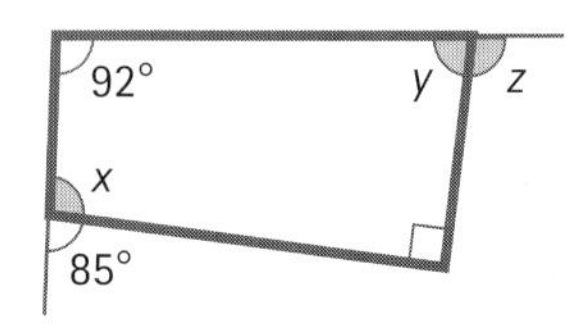

Practice questions

1) Name these polygons.

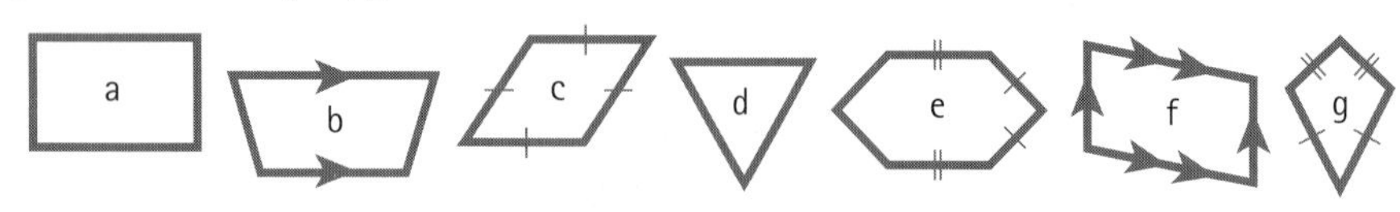

2) Use what you know about the properties of polygons to calculate the missing angles in this sketch of a boat. Give reasons for your calculations.

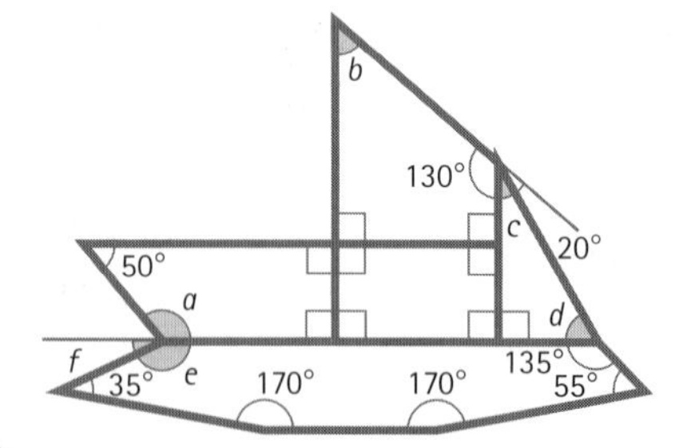

Symmetry, congruence and translation

In this unit you will revise:

- reflective symmetry
- rotational symmetry
- congruence
- translation

Reflective symmetry

A 2D shape has reflective symmetry if a line can be drawn so that:

- when the shape is folded along the line one half fits exactly over the other half
- when a mirror is placed on the line the half shape and its reflection show the whole shape.

REMEMBER The line of symmetry is usually drawn as a dotted line. You can use a mirror to help you find the line of symmetry.

Example 1

This shape has one line of symmetry.

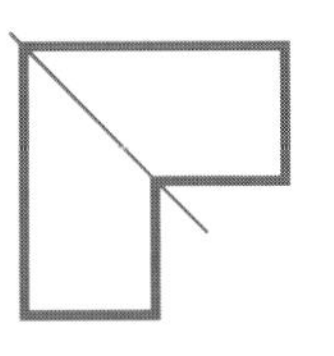
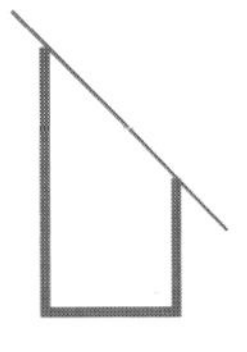

Example 2

This shape has two lines of symmetry.

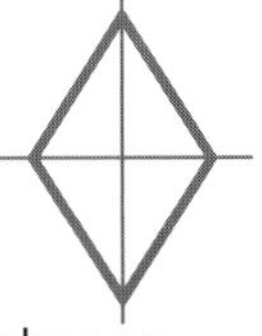
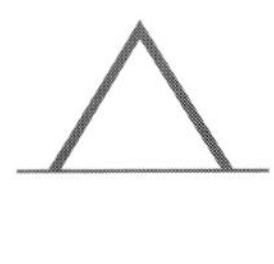
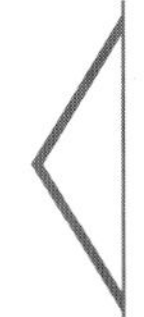

- Draw the lines of symmetry on these shapes.

a) 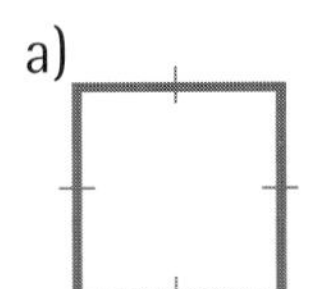

b) 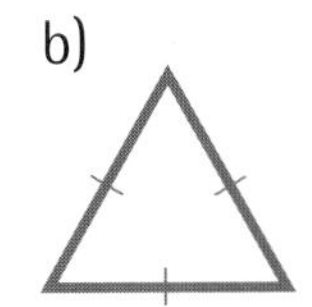

c) 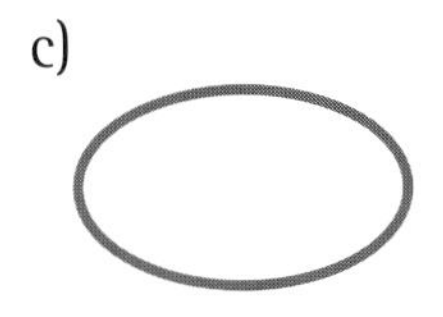

d) 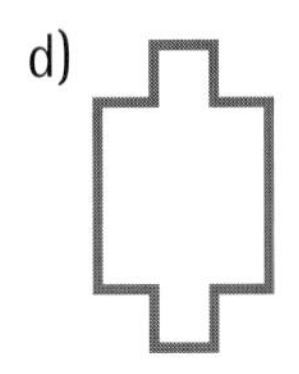

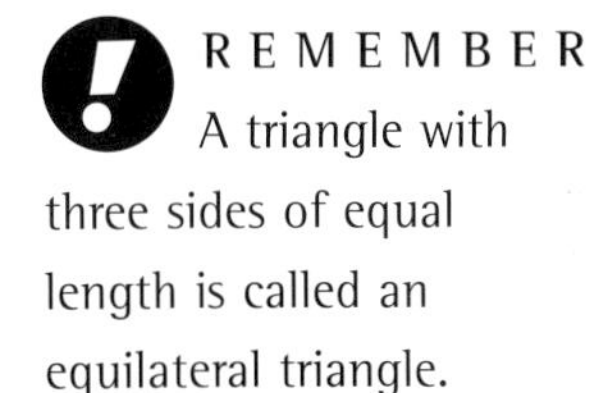

REMEMBER A triangle with three sides of equal length is called an equilateral triangle.

Shape (a) has four sides of equal length, so it is a square.

Shape (b) has three sides of equal length, so it is an equilateral triangle.

The small lines on the sides of shape b show sides of equal length.

REMEMBER It may help to turn the page so you can see the shape in different positions.

Rotational symmetry

A 2D shape has rotational symmetry if it can be rotated and it looks the same in its new position. The number of times a shape can be rotated before it returns to its original position is the order of rotational symmetry.

Example 1

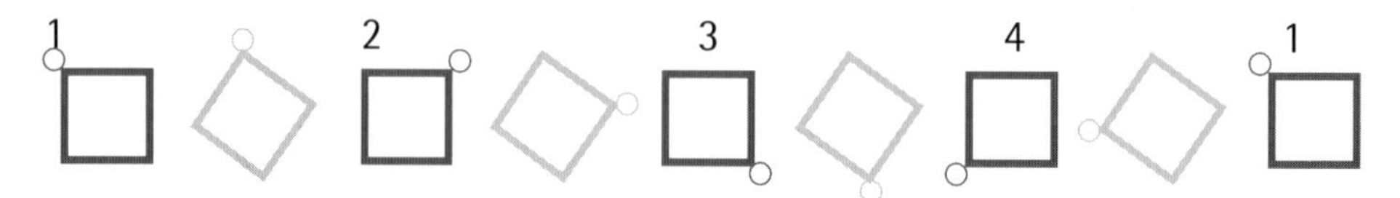

Each time the square is rotated from position 1 to 2 to 3 to 4 it looks the same. The circle shows the new positions of that corner of the square. The square rotates four times before it reaches its original place again. So a square has rotational symmetry of order 4.

Example 2

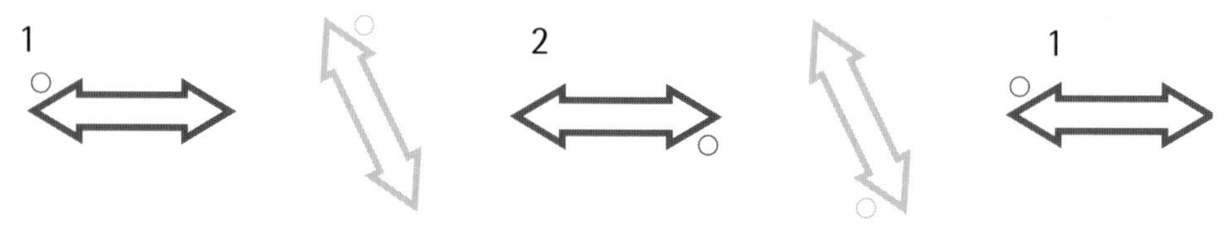

Each time the arrow is rotated from position 1 to position 2 it looks the same. The circle shows the new positions of that end of the arrow. The arrow rotates twice before it returns to its original place. So the arrow has rotational symmetry of order 2.

- Write the order of rotational symmetry for these shapes.

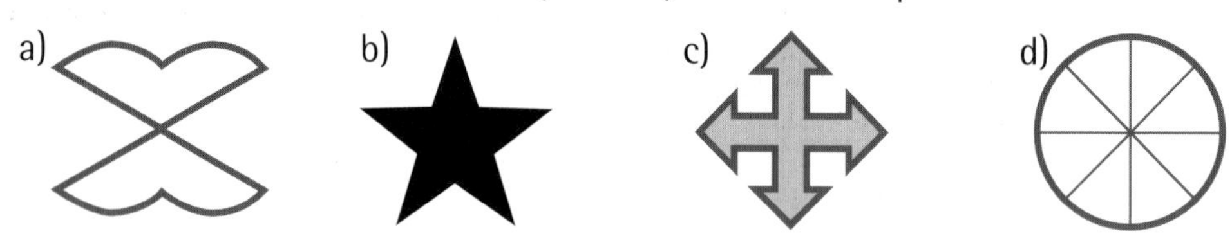

Congruence

2D shapes are congruent if they are both exactly the same shape and size. They may be reflections or rotations of each other.

Example 1

Shapes (a)–(d) are congruent to each other. Shapes (e)–(g) may look congruent at first, but look closer!

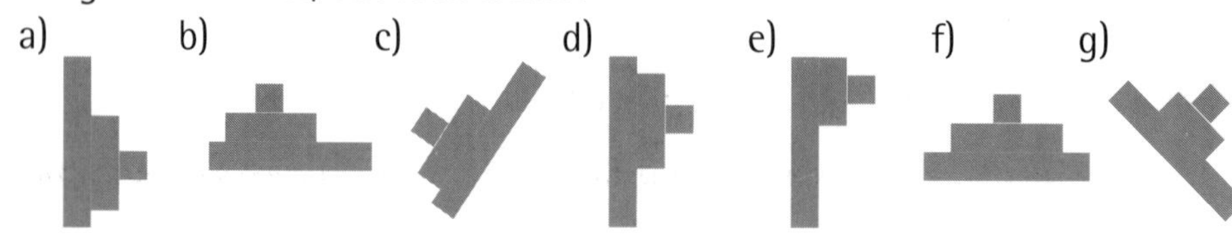

- Which of the three shapes below are congruent?

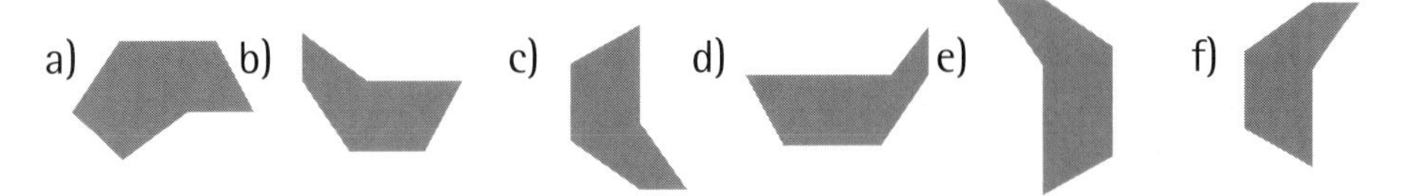

Translation

2D shapes are translated if they move up or down and left or right, remaining exactly the same size and shape, with no rotation, nor reflection.

Example

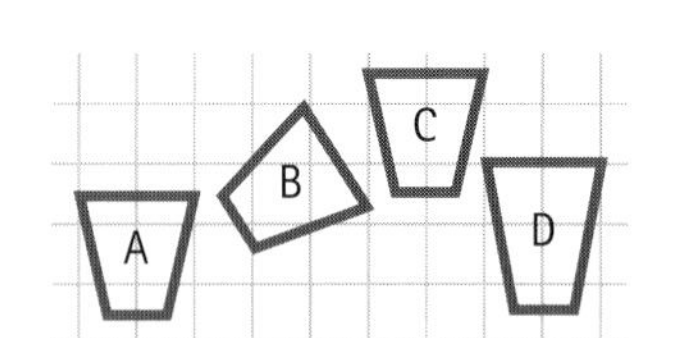

Shape C is a translation of Shape A. All vertices (corners) of Shape A have moved [5 squares right, 2 squares up] to give Shape C.

Shape B is not a translation of Shape A, because it is rotated.

Shape D is not a translation of Shape A, because it is not the same shape.

- Fill in the gap:

  shape E has been translated to shape F [..... squares left, 1 square.....]

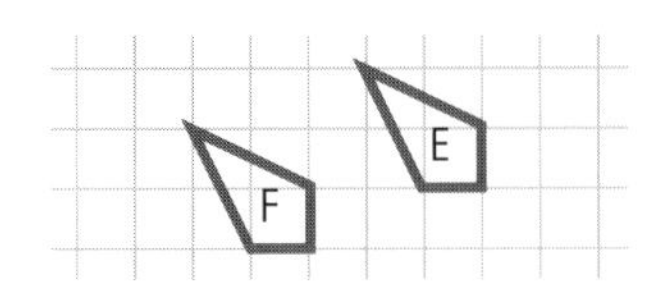

Practice questions

1) Look at the seven patterned square tiles.

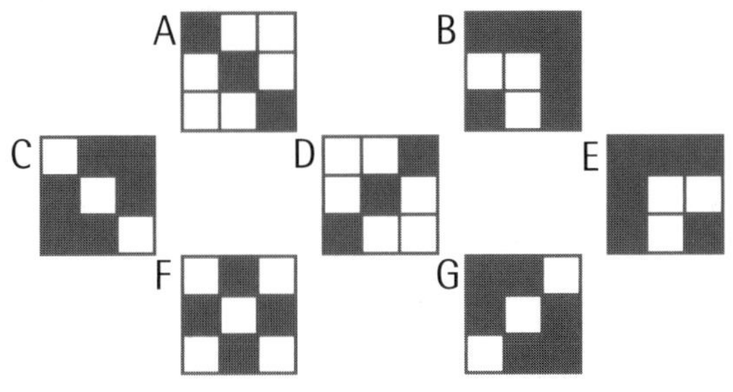

a) Draw lines of symmetry on tiles B, D and F.

b) What is the rotational order of symmetry for tiles A, E and F?

c) Which three pairs of tiles are congruent?

2)

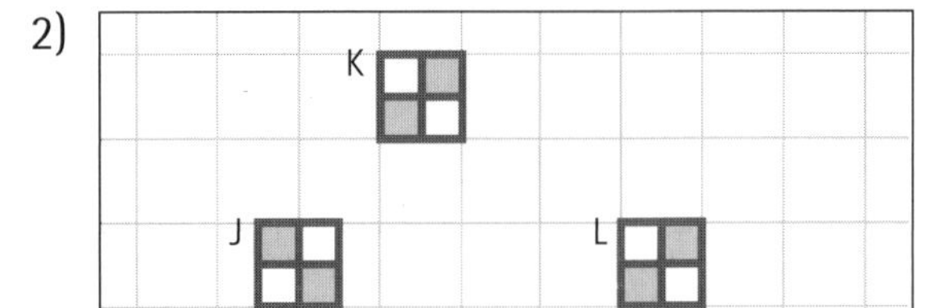

a) Miguel says tile L is a translation of tile J. Is Miguel right?

b) Fill in the gap: tile L has been translated to tile K [3 squares ...., ... squares up]

Enlargement

In this unit you will revise:

- understanding enlargements
- drawing enlargements of shapes

Understanding enlargements

When a shape is enlarged it is made bigger.

The amount a shape is enlarged is measured by a **scale factor**.

Example

Enlarge this rectangle by scale factor 3.

Multiply the length of each of its sides by 3.

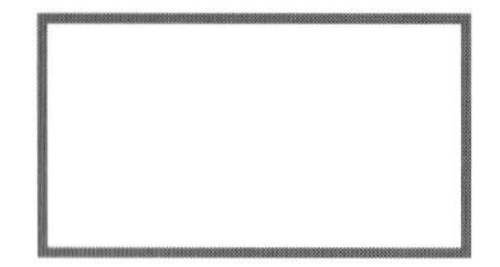

Now each length is 3 times the size and the area is 9 times the size.

Drawing enlargements of shapes

To draw an enlargement, you need to know the point about which to enlarge the shape. This is called the **centre of enlargement**.

Example 1

Enlarge this triangle about the point P by scale factor 2.

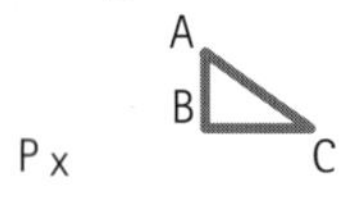

- Draw lines from the centre of enlargement to each vertex.

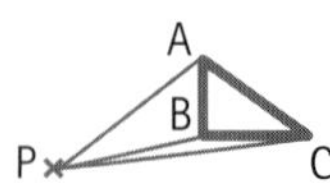

- Measure each line PA, PB and PC and multiply the lengths by 2.

- Draw lines of these lengths from the point of enlargement through the vertices.
- Join the points PA', PB' and PC' to form the enlarged triangle.

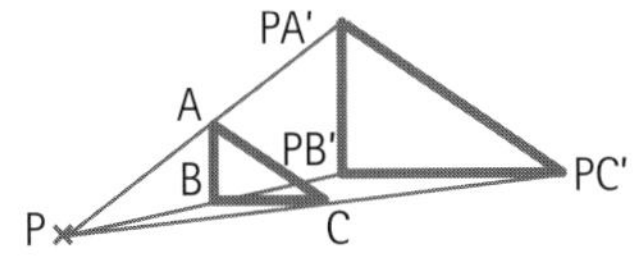

- Draw an enlargement of this rectangle about point O by scale factor 3.

O×

A D
B C

Sometimes shapes are drawn on a coordinate grid. The centre of enlargement is usually the origin (0, 0).

Example 2

Enlarge this rectangle about the origin by scale factor 3.

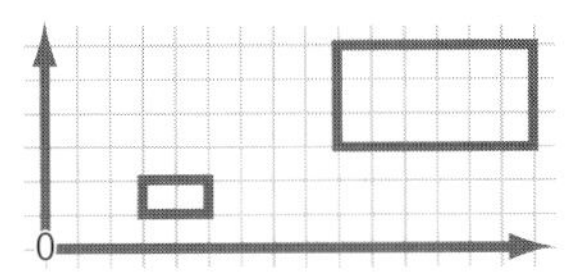

Either follow the steps in example 1.

Or if the centre of enlargement is the origin:

- Find the coordinates of each of the vertices.
  a is (3, 1) b is (5, 1) c is (5, 2) d is (3, 2)
- Multiply each of the pair of co-ordinates by the scale factor.
  a' is (9, 3) b' is (15, 3) c' is (15, 6) d' is (15, 6)
- Plot the new coordinates a', b', c' and d' and join them up to make the enlarged rectangle.

REMEMBER See Coordinates, mappings and line graphs, page 40, to read about plotting coordinates.

Practice questions

Enlarge the shapes below by scale factor 2 about the point O or the origin:

1) O×

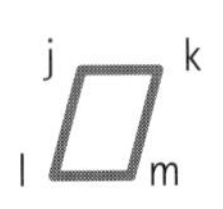

2) O×

u
v
w

3)

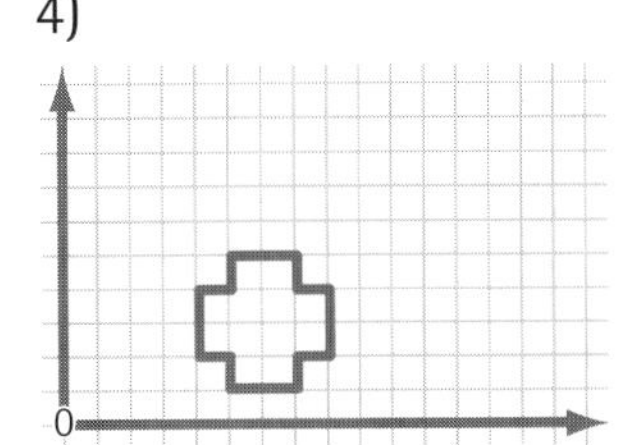

4)

0

Units of measure

> **REMEMBER**
> **Length** is usually measured with a ruler.
> **Capacity** is usually measured with a jug or liquid measures.
> **Weight** is usually measured with scales.
> **Temperature** is usually measured with a thermometer.

In this unit you will revise:

- understanding units of measure
- reading scales accurately
- converting units of measure
- approximate metric and imperial equivalents

Understanding units of measure

Metric units of measure are current standard units of measure. Imperial units of measure are old standard units of measure.

Metric units of length include:
millimetre (mm), centimetre (cm), metre (m), kilometre (km)

Imperial units of length include:
inches (in), feet (ft), yards (yd), miles (m)

Metric units of capacity include:
millilitres (ml), centilitres (cl), litres (l)

Imperial units of capacity include:
pints (pt), gallons (g)

Metric units of weight include:
grams (g), kilograms (kg)

Imperial units of weight include:
ounces (oz), pounds (lb), stones (st)

Metric unit of temperature is:
degrees Celsius or centigrade (°C)

Imperial unit of temperature is:
degrees Fahrenheit (°F)

Reading scales

> **REMEMBER**
> 'cent' means 100 (100 years in a century or 100 legs on a centipede); there are 100 centimetres in 1 metre. 'kilo' means 1000, so there are 1000 metres in 1 kilometre.

Look carefully at the divisions. Decide how many units each division represents.

Example 1

On this ruler each division is 2 mm.

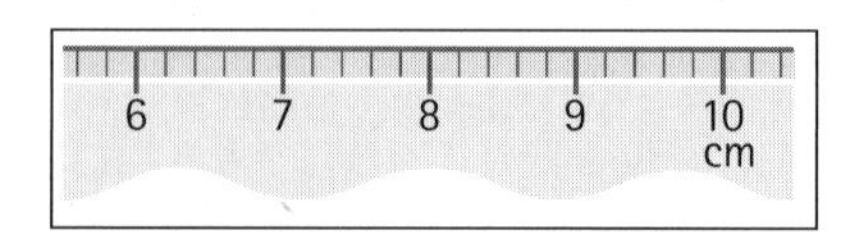

Example 2

On this cup each division is 250 ml.

• Write down how many °C the divisions on this thermometer represent and what temperature it shows.

Converting units of measure

Converting units of measure is easier if you know the metric unit conversions.

Metric units

Length	10 mm = 1cm	100 cm = 1m	1000 m = 1km
Capacity	100 cl = 1 litre	1000 ml = 1 litre	
Weight	1000 g = 1 kg		

Imperial units

Length 12 inches = 1 foot   3 feet = 1 yard   1760 yards = 1 mile
Capacity 8 pints = 1 gallon
Weight 16 ounces = 1 pound   14 pounds = 1 stone

To change one unit of metric measure to another or
one unit of imperial measure to another:

- to change from a large measurement to a small measurement, multiply, x
  to change from a small measurement to a large measurement, divide, ÷
- remember the right conversion!

REMEMBER Whole numbers, page 8, covers multiplying and dividing whole numbers by 10, 100, 1000. Decimals, page 10, covers multiplying and dividing decimals by 10, 100.

Example 1

Change 7.4 m to cm:
100 cm = 1 m   so x 100
7.4 m = 7.4 x 100 = 740 cm

Example 2

Change 3000 g to kg:
1000 g = 1 kg   so ÷ 1000
3000 g = 3000 ÷ 1000 = 3 kg

Example 3

How many ounces are there in 20 pounds?
16 ounces = 1 pound so x 16
20 pounds = 20 x 16 = 320 ounces

• a) How many cm make 378 mm?   b) Change 6 gallons into pints.

changing:
- m → cm
  large → small   x
- g → kg
  small → large   ÷
- pounds → ounces
  large → small   x

Approximate metric and imperial equivalents

Length	1 in is about 2.5 cm 1 yd is about 90 cm	1 ft is about 30 cm 1 mile is about 1.6km
Capacity	1 pint is about 0.5 litres	1 gallon is about 4.5 litres
Weight	1 ounce is about 30 grams 1 stone is about 6.5 kilograms	1 pound is about 450 grams

To change a unit of imperial measure to a unit of metric measure:

- remember the right conversion
- multiply the imperial quantity by the metric conversion value.

Example 1

Change 4 pints into litres.
The conversion is 1 pint is about 0.5 litres
4 pints = 4 x 0.5 = 2 litres (approx)

REMEMBER Make a list of all the conversions you need to remember. Spend a few minutes every day testing yourself on them and before long you will find that they will have stuck in your memory

To change a unit of metric measure to a unit of imperial measure:

- remember the right conversion
- divide the metric quantity by the imperial conversion value.

Example 2

Change 270 grams into ounces.
The conversion is 30 g = 1 oz
270 grams = 270 ÷ 30 = 9 oz

Practice questions

1) Look carefully at the scales on this equipment for measuring.

Notice that the scales do not always start at zero.

a) Approximately how long is the pencil from end to tip?

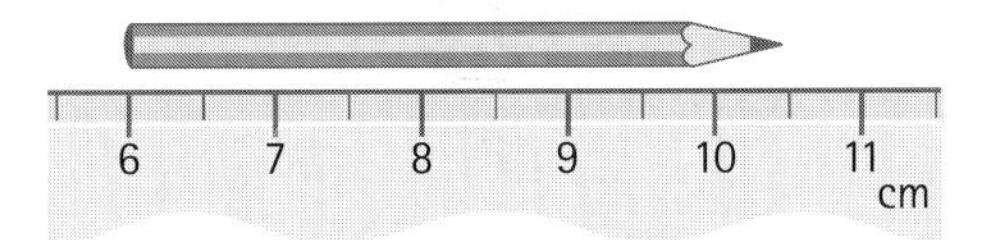

b) Approximately how much water is there in the jar?

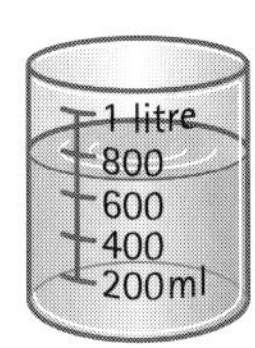

c) Approximately what temperature is it?

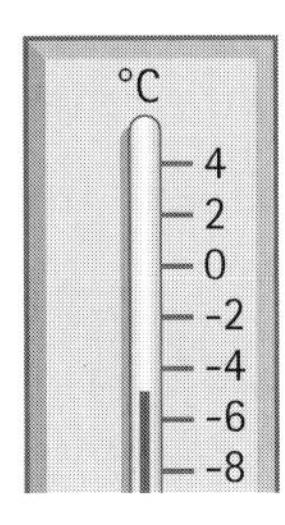

2) a) Anne's foot is 25cm long. How many inches is that?

b) Chris's waist is 95 cm round. How many metres is that?

c) Henry's neck is 33 cm round. How many millimetres is that?

d) Dora's height is 4.5 feet. How tall is that in centimetres?

3) Sarah runs 3 miles, Mark runs 5 km.

Sarah says she has run further. Is she right?

FactZONE

Time

60 seconds = 1 minute
7 days = 1 week
365 days = 1 year
60 minutes = 1 hour
28, 29, 30 or 31 days = 1 month
366 days = 1 leap year
24 hours = 1 day
12 months = 1 year
52 weeks = 1 year

To change minutes into seconds multiply by 60.

The clock

Time may be measured in the 12-hour clock.

The day is divided into am, midnight to noon, e.g. 11.15 am, 8.43am
pm, noon to midnight, e.g. 11.15 pm, 8.43pm

Time may be measured in the 24-hour clock.
Then 11.15am is written as 1115 hours and 8.43am is written as 0843 hours
11.15pm is written as 2315 hours and 8.43pm is written as 2015 hours

The 24-hour clock is always written with four figures.

To change from the 12-hour clock to the 24-hour clock, add 12 to the hours.
To change from the 24-hour clock to the 12-hour clock, take away 12 hours.

How to use timetables

Timetables are usually written in the 24-hour clock.

Look carefully at this bus timetable.

East St	0845	0915	0945
North St	0858	0928	...
West St	0910	0940	1006
	This column shows what time the first bus arrives at each stop	This column shows what time the second bus arrives at each stop	This column shows what time the last bus arrives at each stop.

Perimeter

In this unit you will revise:

- understanding perimeter
- finding perimeters of simple shapes
- finding perimeters of circles

Understanding perimeter

The perimeter of a shape is the distance all the way round the outside.

So the perimeter of this rectangle is 7 + 1.5 + 7 + 1.5 = 17 cm

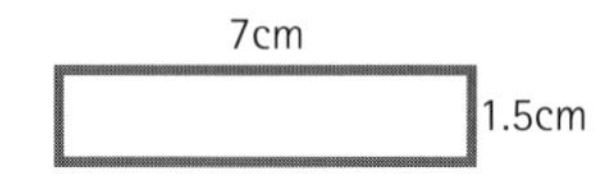

Finding perimeters of simple shapes

Sometimes finding the perimeter of a shape is simple.

The perimeter of this diamond is 2 + 2 + 2 + 2 = 8 cm

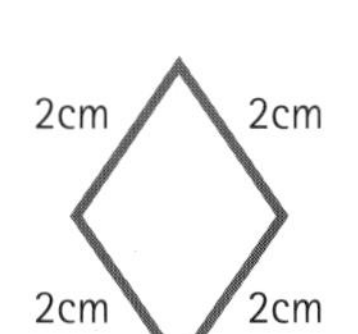

Even if the shape is made up of smaller shapes, you still count the distance round the outside. Ignore the lines inside the shape.

The perimeter of this triangle is 2 + 2 + 2 = 6 cm

made up of these

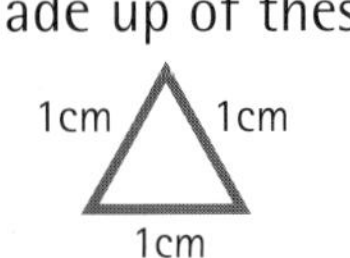

REMEMBER When you are finding the perimeter of a shape, make sure you include all the sides.

Sometimes the shape is irregular.

The perimeter of this shape is 0.5 + 6 + 5.5 + 4 + 2 + 4 = 22 cm

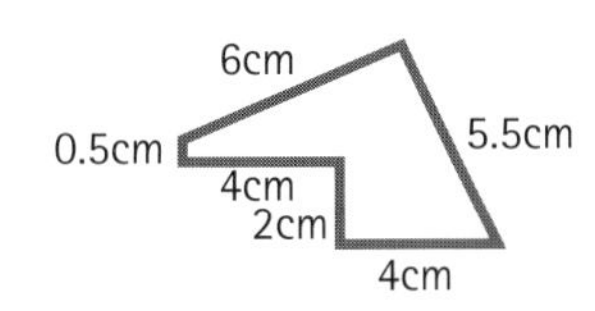

Finding perimeters of circles

The perimeter of a circle is called its circumference.

To find the perimeter of a circle you need to know the diameter or radius.

A diameter, d, is a straight line from any point on the circle through the centre to any other point on the circle.

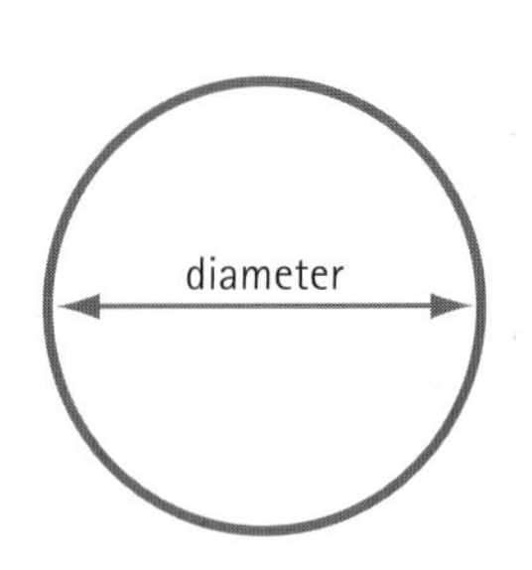

A radius, r, is any straight line from a point on the circle to the centre.

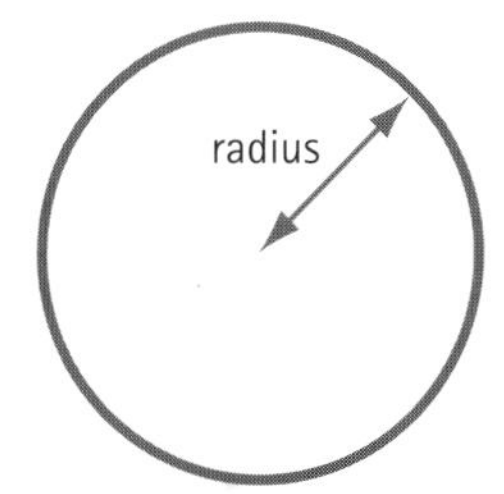

So the diameter = 2 x radius $d = 2r$
To find the circumference, C, use the formula C = 3.14 x diameter or C = 3.14 x 2 x radius

REMEMBER See Rounding and approximating, page 22, to read about rounding to the nearest whole number.

The ratio of the circumference to the diameter is rounded to 3.14.
A closer value is 3.141 592 653 589 793 238 462 643 383 279 5...

In fact, the decimal fraction goes on for ever, without repeating, so we can never use the entire number. This is why the ratio is called by the Greek letter, pi, written π.

Strictly, circumference = π x diameter or π x 2 x radius.

If you are asked for an approximation, '3 x diameter' will do!

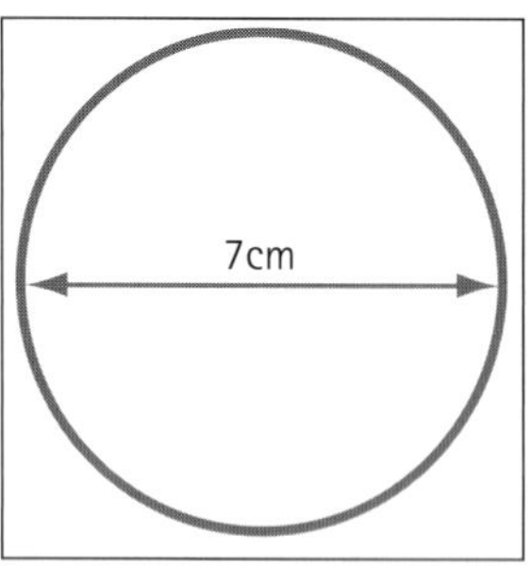

Example 1

Find the circumference of this circle.

diameter = 7 cm
circumference = 3.14 x 7 = 21.98 cm

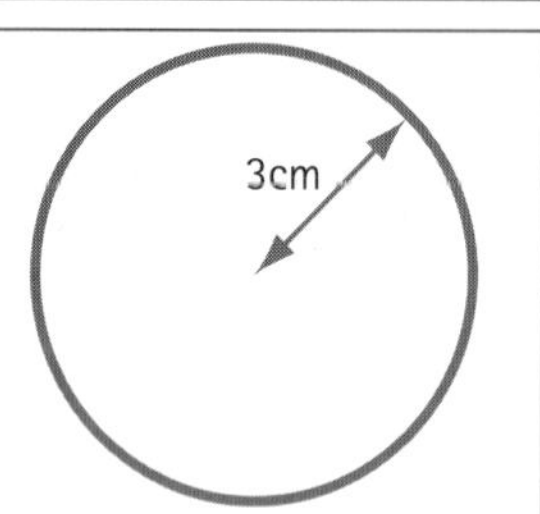

Example 2

Find the circumference of this circle.

radius = 3 cm, so diameter = 2 x 3 = 6 cm
circumference = 3.14 x 6 = 18.84 cm

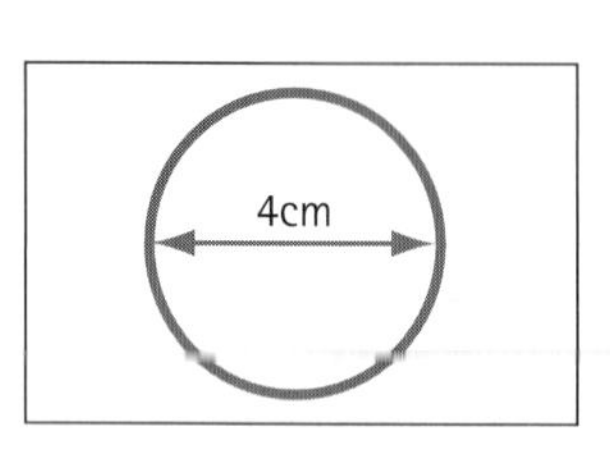

Example 3

Find an approximation for the circumference of this circle.

diameter = 4 cm
circumference = 3 x 4 = 12 cm

Practice questions

1) Find the perimeters of the four shapes below.

a)

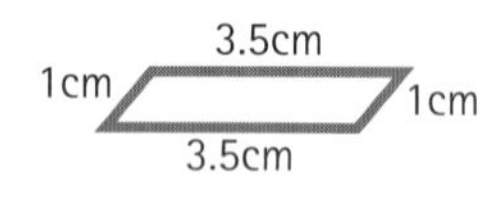

b)

where

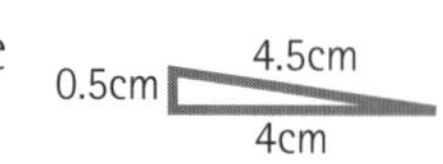

2) Find the circumferences of these circles:

(i) using the formula
circumference = 3.14 x diameter

(ii) using the approximation
circumference = 3 x diameter

a) b) c)

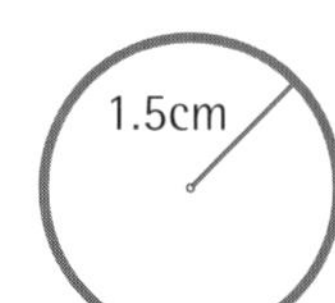

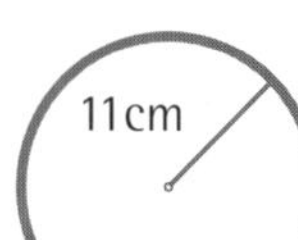

In this unit you will revise:

- understanding area
- finding the area of simple shapes
- using formulae to find area

Understanding area

The area of a 2D shape is the amount of space covered by the shape.

The units of area are called 'square' units and might be mm^2, cm^2, m^2 or km^2.

Finding the area of simple shapes

This is 1 cm^2

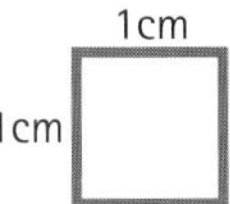

This is 2 cm^2

This is 3 cm^2

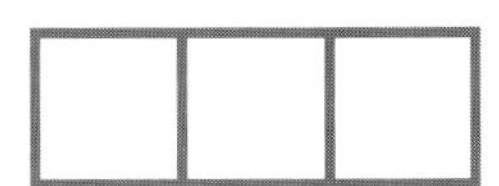

You can find areas by counting squares.

Example 1

The area of this rectangle is 6 units squared.

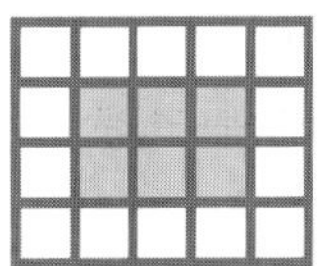

Example 2

The area of this triangle is 6 x □ and 4 x ◺
But (4 x ◺ = 2 x □) so the area is 6 x □ + 2 x □
So the area is 8 units squared.

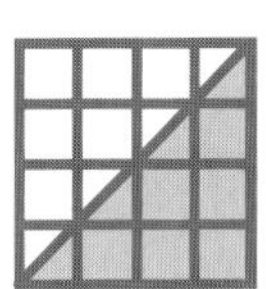

- What is the area of this shape?

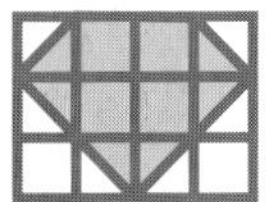

REMEMBER See Using formulae, page 31, to read about using formulae.

Using formulae to find area

Look again closely at example 1 and example 2 above.

The area of the rectangle could be found as base length x height.

base length x height = 3 x 2 = 6 units squared

So you write the formula for area of a rectangle as
$A = b \times h$

and the area of the right-angled triangle can be found as
$A = \frac{1}{2}$ x base length x height

$\frac{1}{2}$ x base length x height = $\frac{1}{2} \times 4 \times 4 = 8$ units squared

So you write the formula for the area of a right-angled triangle as
$A = \frac{1}{2} \times b \times h$

FactZONE

Formulae for the areas of shapes

area of a square = length l x length l
$= l \times l$
$= l^2$

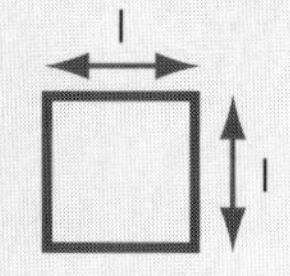

area of triangle = $\frac{1}{2}$ x length b x length h
$= \frac{1}{2} \times b \times h$
$= \frac{1}{2}bh$

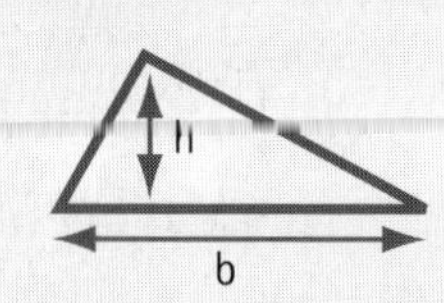

area of a parallelogram = length b x height h
$= b \times h$
$= bh$

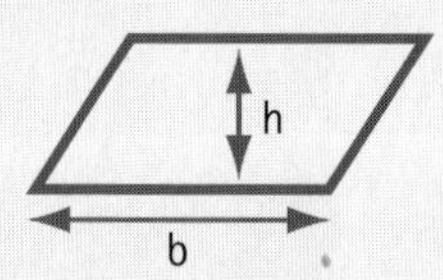

area of a trapezium = $\frac{1}{2}$(length a + length b) x height h
$= \frac{1}{2}(a + b)h$

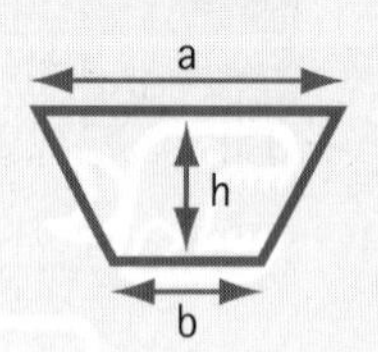

area of a circle = π x radius x radius
$= \pi \times r \times r$
$= \pi r^2$

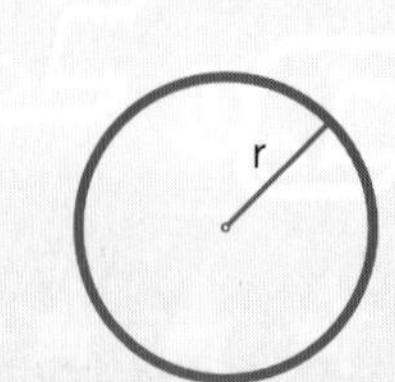

- Use the correct formulae to find the areas of these shapes.

a) square

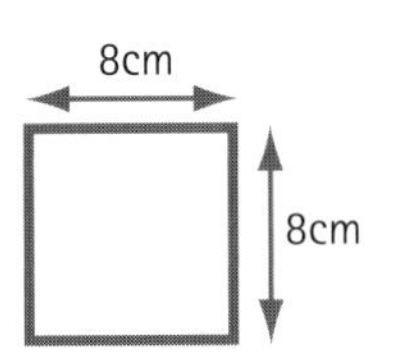

b) trapezium

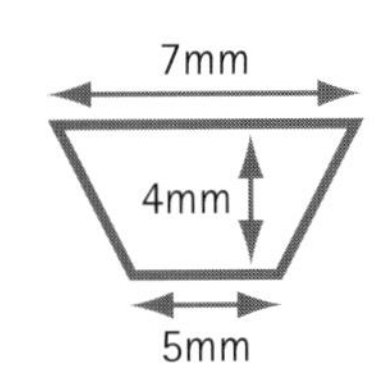

c) circle

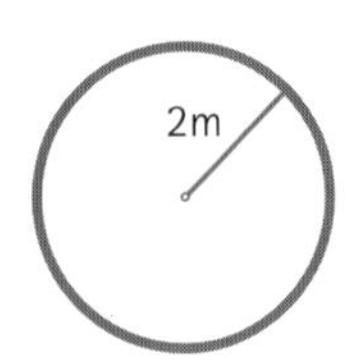

REMEMBER See Perimeter, page 67 and 68, to read about radius and π.

Practice questions

Find the areas of the shapes below by either counting squares or using the formulae.

1)

2)

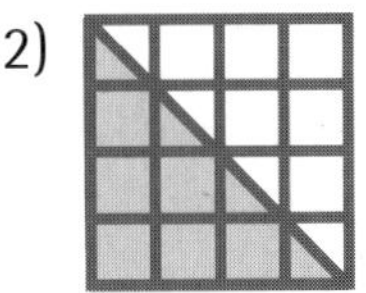

3)

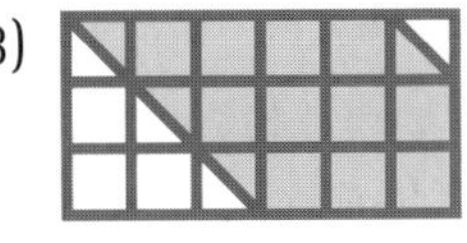

4)

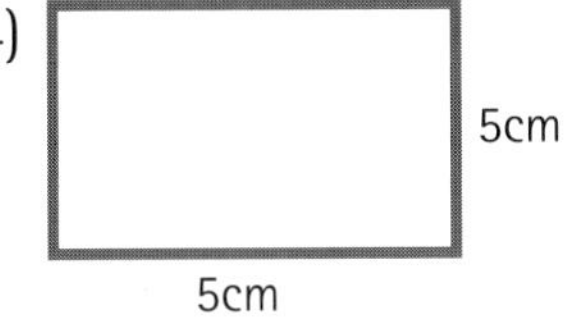

5)

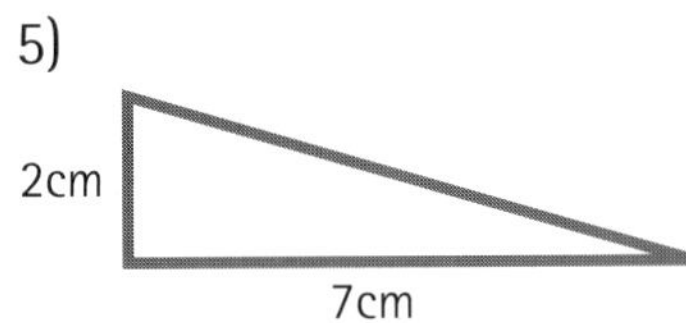

6)

7)

6cm
3cm
2cm

8)

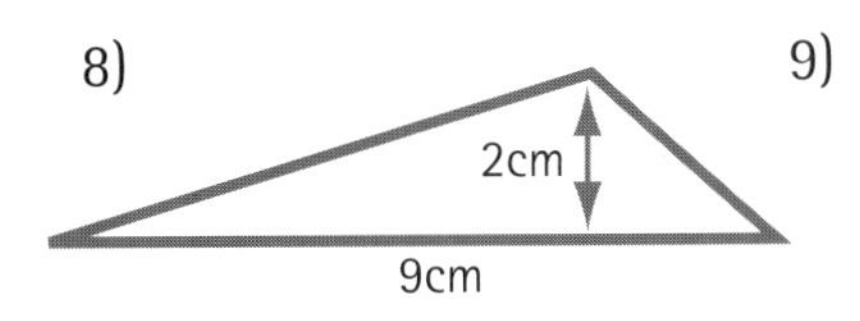

9)

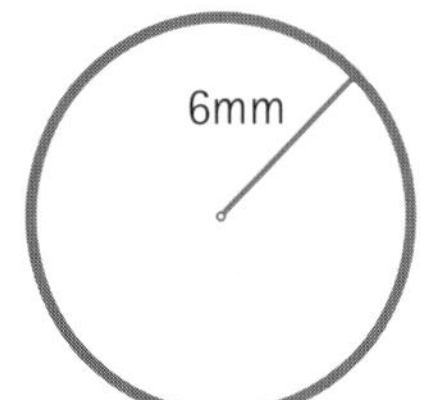

10)

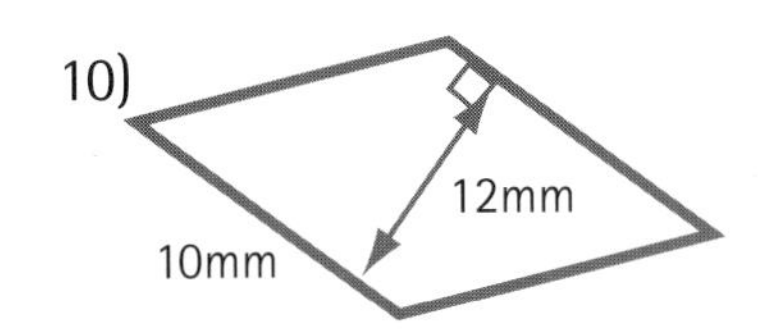

11)

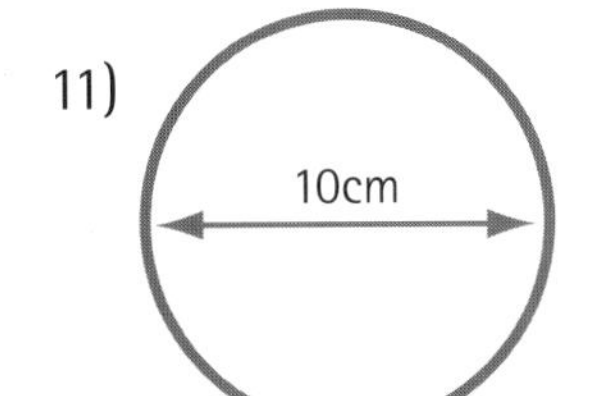

12)

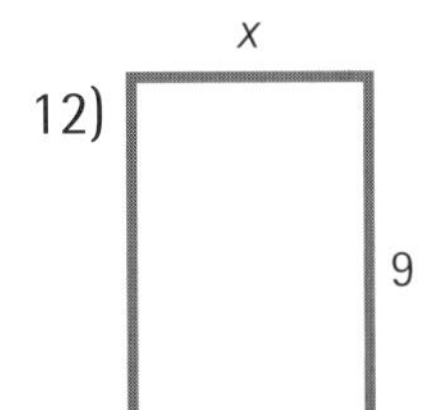

In this unit you will revise:

- understanding volume
- finding the volume of simple solids
- using formulae to find volume

Understanding volume

The volume of a solid is the amount of space it takes up.

Units of volume are 'cubic' units and might be mm^3, cm^3, m^3 or km^3.

Finding the volume of simple solids

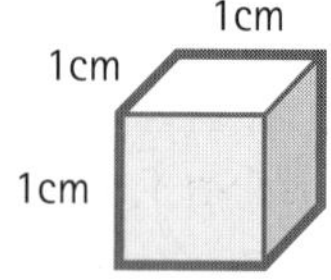

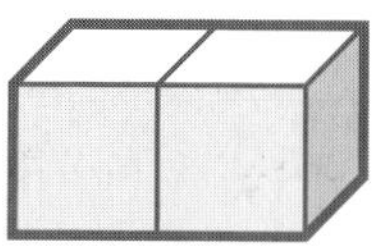

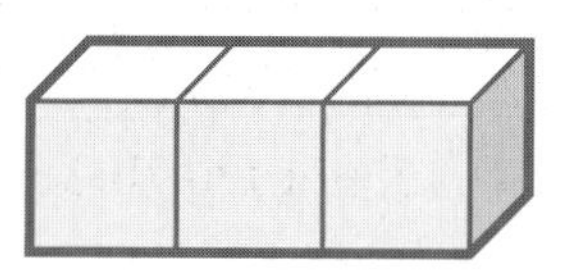

This is 1 cm^3 This is 2 cm^3 This is 3 cm^3

You can find volume by counting cubes.

Example 1

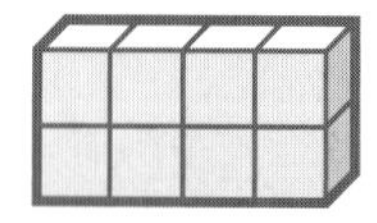

The volume of this cuboid is 8 units cubed.

Example 2

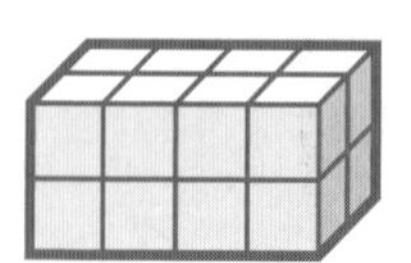

The volume of this cuboid is 16 units cubed.

Using formulae to find volume

Take another look at example 1 and example 2.

The volume of each cuboid could be found by multiplying. The formula is:
volume = length x width x breadth:

in example 1 length x width x breadth = 4 x 2 x 1 = 8 units cubed

in example 2 length x width x breadth = 4 x 2 x 2 = 16 units cubed

so you write the formulae for the volume of a cuboid as $V = l \times w \times b$

Practice questions

Find the volumes of these shapes, by either counting cubes or using formulae.

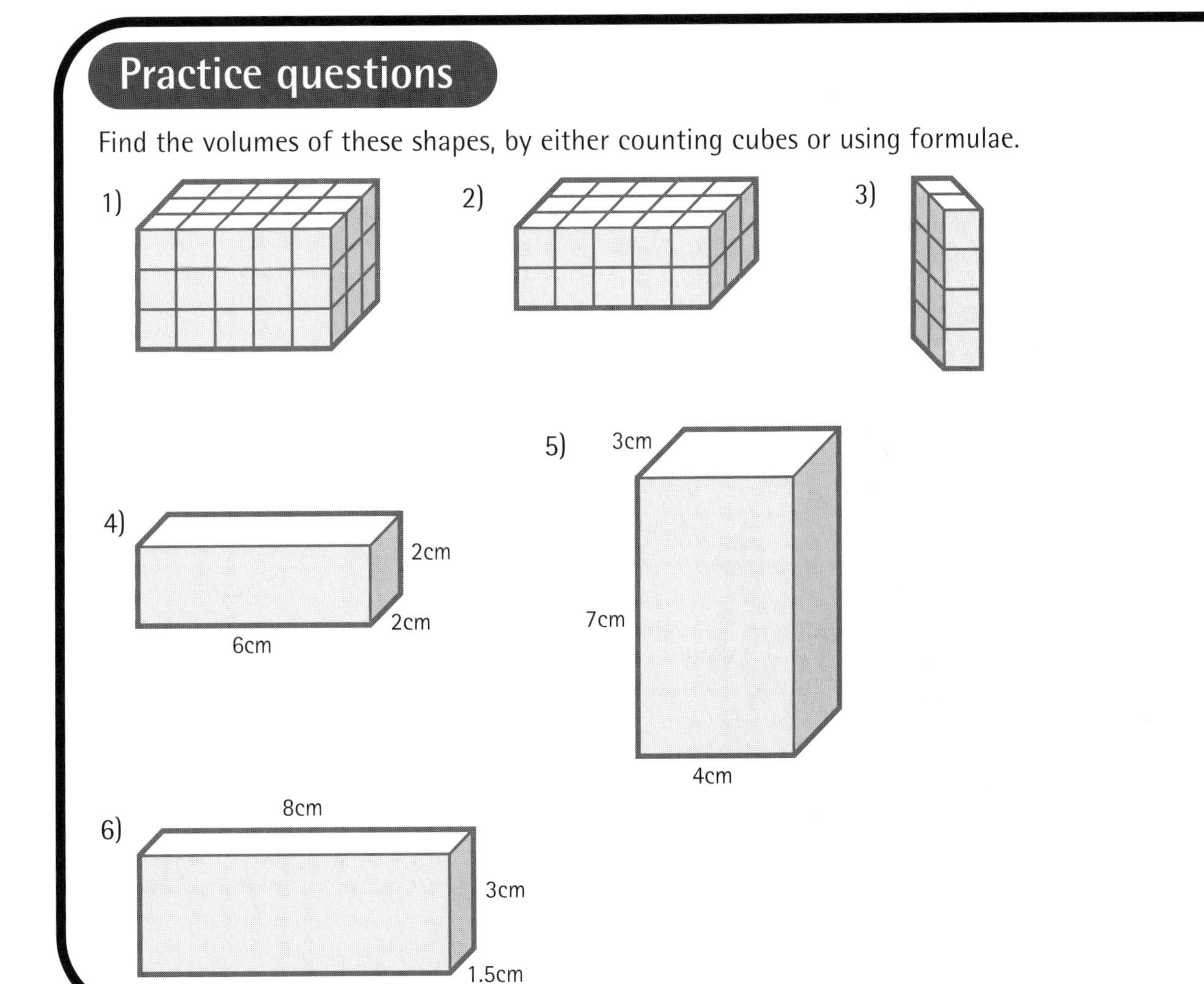

Volume

Representing and interpreting data

In this unit you will revise:

- tallies, frequency tables and bar charts
- drawing and interpreting line graphs
- interpreting pie charts

Tallies, frequency tables and bar charts

Data that is collected but is not put in any order is called raw data.

Example 1

Waterman United football team's scores this season are:

0 5 0 2 1 2 1 1 1 1 2 3

0 2 4 1 1 0 3 1 2 0 4 0

You can use a tally chart to draw up a frequency distribution.

Number of goals	Tally	Frequency
0	~~||||~~ |	6
1	~~||||~~ |||	8
2	~~||||~~	5
3	||	2
4	||	2
5	|	1

Goals scored by Waterman United

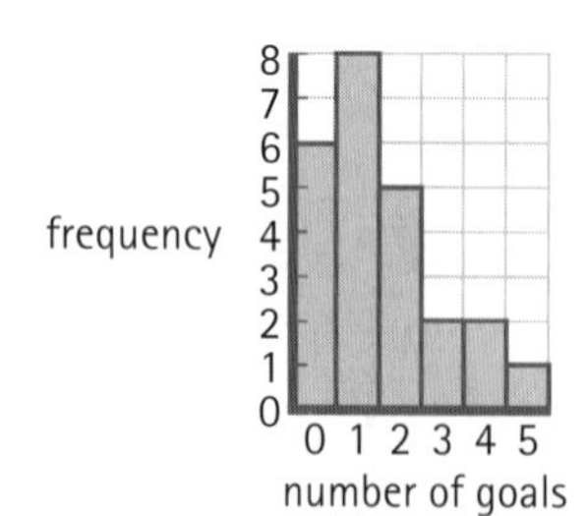

You can use a frequency chart to display the data on a bar chart.

The bar chart shows the numbers of goals scored by Waterman United. You can see that:

- each axis is labelled
- each bar is the same width
- the length of each bar corresponds to the frequency for that score
- the graph has a title.

Large amounts of data spread over a large range can be grouped.

Example 2

Here are the scores of 33 students in a mental maths test.

24	24	22	20	16	20	12	17	20	21	15
12	20	20	20	9	16	20	8	17	21	24
19	11	15	9	11	16	25	21	9	22	25

The scores range from 8 to 25. This is a large range, so order the information into class intervals of width 5 marks.

Class interval	Tally	Frequency
0–5		0
6–10	////	4
11–15	𝍸	5
16–20	𝍸 𝍸 ////	14
21–25	𝍸 𝍸	10

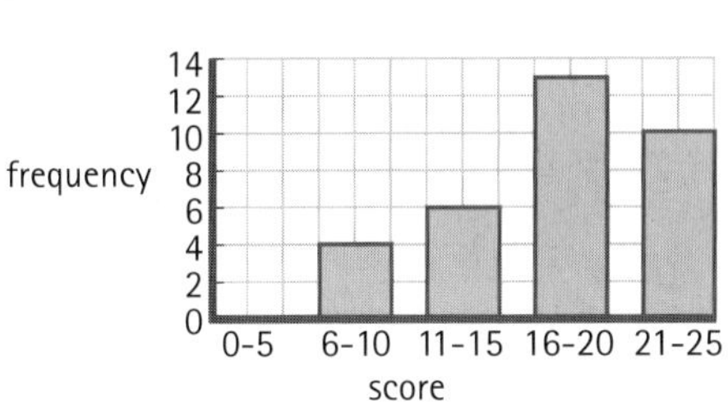

Drawing and interpreting line graphs

Line graphs can be used to display data.

Example 1

In Osney newsagents the sales of Mr Cooly ice-creams each month from March to September are:

Month	Mar	Apr	May	Jun	Jul	Aug	Sep
Sales	100	130	170	210	240	250	210

Mr Cooly's ice-cream sales

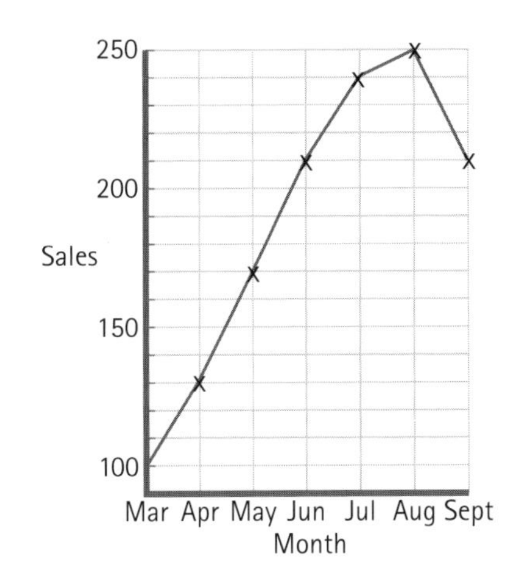

You can see that:

- each axis is labelled
- for sales each division represents 10 sales
- for months each division represents 1 month
- points are marked with small crosses and joined by straight lines
- the graph is drawn with a sharp pencil
- the graph has a title.

Representing and interpreting data

Sales of Mr Icy lollies

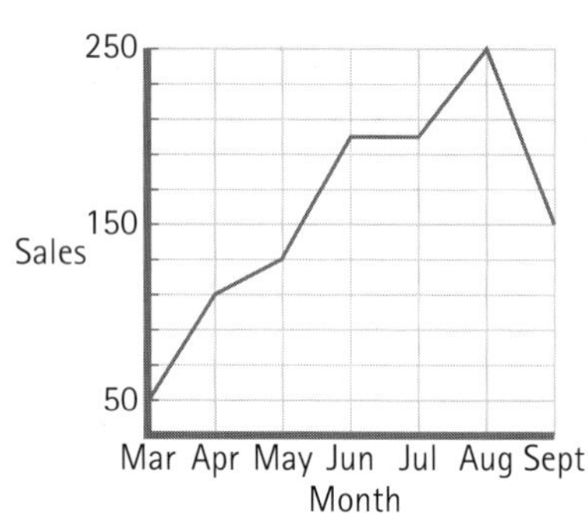

Example 2

Reading from the graph, in Osney newsagents the sales of Mr Icy lollies each month from March to September are:

Month	Mar	Apr	May	Jun	Jul	Aug	Sep
Sales	50	110	130	200	200	250	150

Conversion graphs are line graphs that can be used to convert:

- imperial units of measure to metric units of measure and back again
- currency.

Banks' exchange rates tell you how much foreign currency you get for £1.

REMEMBER See Units of measure, page 62, to read about understanding units of measure.

Example 3

The exchange rate between the pound and the Euro is £1 to 70 Euros, so for every pound you would get 70 Euros.

To work out how many Euros you get for £10:

£10 = £1 x 10

so £10 = E 70 x 10 = E 700

How would you work out how many pounds you get for E 175?

The calculation is not quite so easy, but a conversion graph helps.

From the graph: E 175 = £2.50

Conversion rate between Euros and pounds

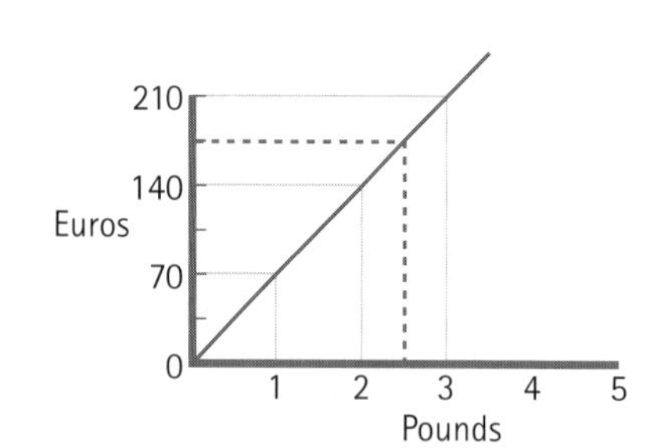

You can also see:
70 Euros = £1

140 Euros = £2

210 Euros = £3

REMEMBER See Fractions, page 14. To read about understanding fractions.

Interpreting pie charts

You can also display data in a pie chart. A pie chart is a diagram, which looks like a circular pie cut into portions. To interpret pie charts, you need to understand fractions.

Example

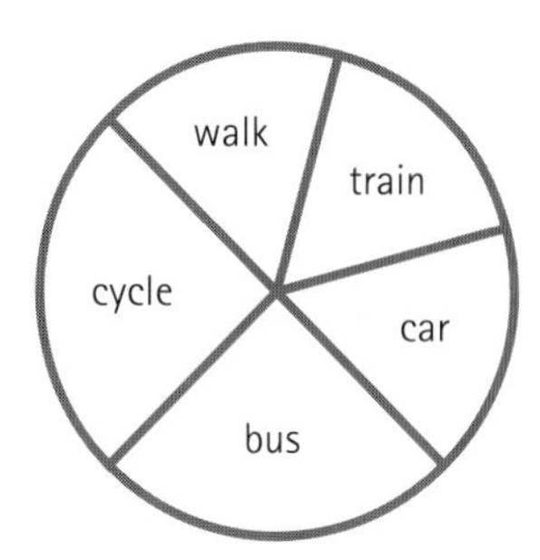

The pie chart shows the modes of transport used by Year 9 pupils at Osney High to get to school. Look at the fractions of the circle taken up by each portion. Notice: the same number of pupils travel to school by foot, train or car; the same number of pupils travel to school by bicycle or bus.

Look at the fractions:

$\frac{1}{6}$ of Year 9 walk
$\frac{1}{6}$ of Year 9 travel by train
$\frac{1}{6}$ of Year 9 travel by car
$\frac{1}{4}$ of Year 9 travel by bus
$\frac{1}{4}$ of Year 9 cycle.

30 Year 9 pupils have bus passes.

So 30 Year 9 pupils travel to school by bus.
30 pupils make up $\frac{1}{4}$ of Year 9.

So there are 30 x 4 = 120 pupils in Year 9.

- How many pupils: a) travel to school by bus? b) walk to school?

Practice questions

Year 9 students at Osney High vote for a charity to support each year.
The number of votes each charity received

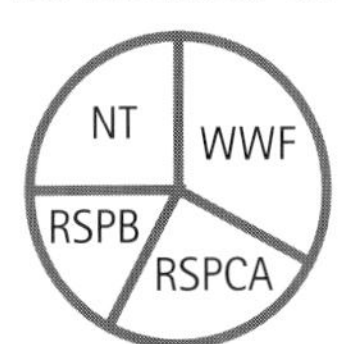

Types of events pupils held to raise money

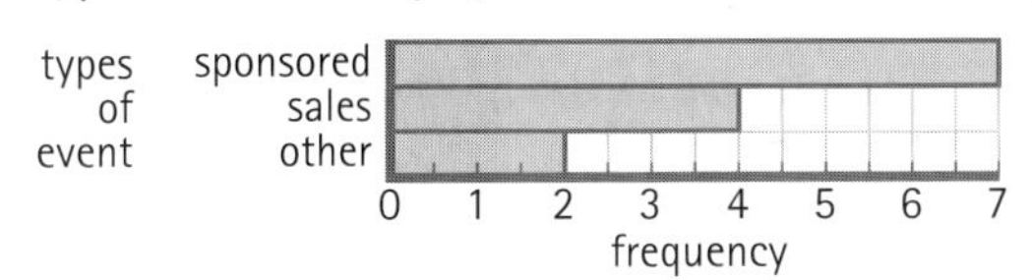

Amount of money raised each half-term

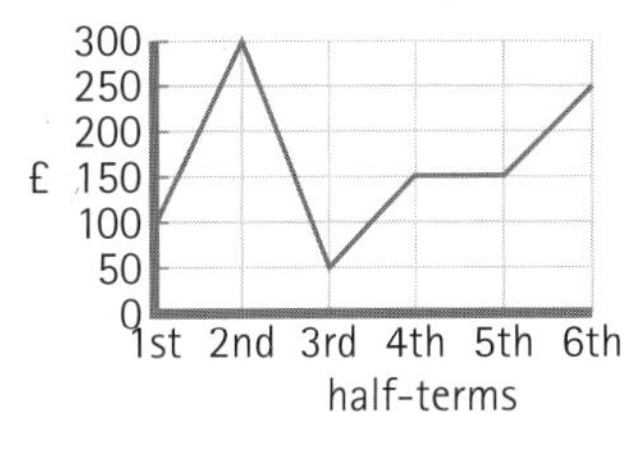

1) Which charity was chosen?
2) What fraction of Year 9 voted for the RSPCA?
3) If 108 pupils voted, how many voted for NT (the National Trust)?
4) How many sponsored events did Year 9 hold?
5) Which event did they hold four times?
6) How much money did Year 9 raise in the fifth half-term?
7) How much money did Year 9 raise in total?

Representing and interpretting data

 Comparing distributions

In this unit you will revise:

- averages
- comparing data using diagrams
- interpreting scatter diagrams

Averages

There are three important types of average – mode, median and mean.

Example 1

Look at this information about the performance of Wickers hockey team.

Last season Wickers hockey team played 13 matches and the numbers of goals scored in their matches have been listed.

6 10 4 2 4 1 7 4 1 4 4 1 4

This season Wickers hockey team played 12 matches and the numbers of goals scored in their matches have been entered on this bar chart.

Notice that the graph shows that Wickers hockey team scored:

Goals scored this season

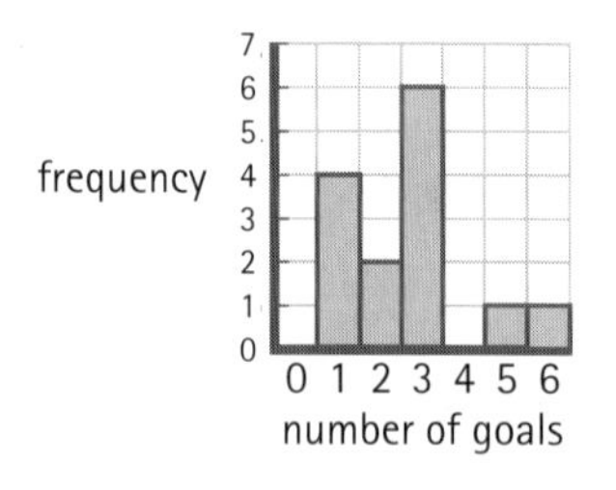

- 1 goal in four matches
- 2 goals in two matches
- 3 goals in six matches
- 5 goals in one match
- 6 goals in one match.

REMEMBER The mode is sometimes called the most popular value, as it occurs more than any other.

- **Mode – the quantity that occurs most often in a frequency distribution**

For last season, 4 occurs most often in the frequency distribution, as listed, so the modal number of goals is 4. For this season, 3 occurs most often in the frequency distribution, as shown on the bar chart, so the modal number of goals is 3.

- **Median – the middle quantity when all quantities are placed in order of size**

For last season, placing the scores in numerical order, as listed:

1 1 1 2 4 4 4 4 4 4 6 7 10

There are 13 quantities, so the middle quantity is the 7th quantity.

The median number of goals is 4.

For this season, placing the scores in numerical order, as shown on the bar chart:

1 1 1 1 2 2 3 3 3 3 3 3 5 6

There are 14 quantities, so the middle quantity falls between 3 and 3. So the median number of goals is 3.

- **Mean – the total of all quantities divided by the number of quantities**

For last season, the total of all quantities (goals) was

$6 + 10 + 4 + 2 + 4 + 1 + 7 + 4 + 1 + 4 + 4 + 1 + 4 = 52$

The number of quantities (goal scores) was 13, so the mean number of goals is $52 \div 13 = 4$.

For this season, the total of all quantities (goals) was

$1 + 1 + 1 + 1 + 2 + 2 + 3 + 3 + 3 + 3 + 3 + 3 + 5 + 6 = 37$

The number of quantities (goal scores) was 14, so the mean number of goals is $37 \div 14 = 2.64$

Note: It is impossible to score 2.64 goals, so you say 3, to the nearest goal!

Finding the **range** can help you understand an average.

range = highest value – lowest value

Last season, the range of scores by Wickers hockey team was

$10 - 1 = 9$

This season, the range of scores by Wickers hockey team was

$6 - 1 = 5$

Therefore Wickers hockey team had a lower range this season, showing that their performance was more consistent.

Example 2

Mary, Nigel, Amy and Oliver were training for a swimming relay race. At their final four training sessions the coach recorded the time it took, to the nearest second, for each of them to swim a length of the pool. Here are the results.

Mary	Nigel	Amy	Oliver
39 38 36 33	31 30 32 30	30 35 29 26	40 34 33 32

- Answer these questions.

a) Look at the 16 times for the whole team. What are the median, mode and mean times for the whole team?

b) The swimming coach orders the team for the relay race according to each team member's time range. She decides that the swimmer with the smallest time range will swim first, then the swimmer with the next smallest range and so on, with the last swimmer being the team member with the greatest time range. What order does the team swim, in the race?

Comparing data using diagrams

Always look carefully at diagrams and try to understand what they are telling you before making any comparisons.

There are 24 students in Class M2. The pie chart shows the sports played by pupils in Class M2.

Sports played by pupils in Class M2

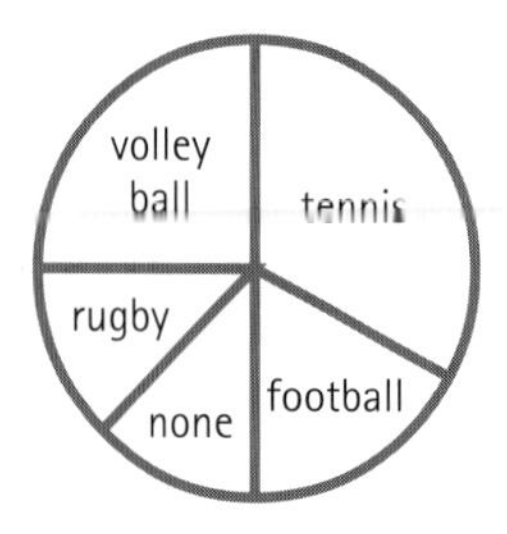

There are 20 students in Class N2. The bar chart shows the types of musical instruments played by pupils in Class N2.

Musical instruments played by pupils in Class N2

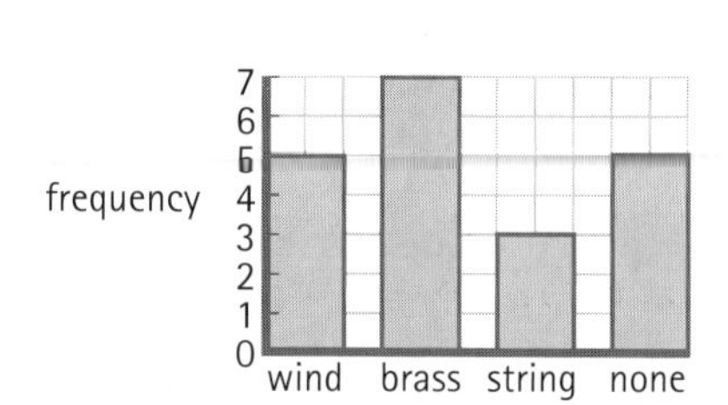

- Are there more or fewer volleyball players in Class M2 than there are Brass players in Class N2? Show your working.

Interpreting scatter diagrams

Scatter diagrams allow you to compare two sets of data on the same diagram.

Example

In April, Noho City Council conducted a survey for two weeks, into how temperatures affect the use of leisure facilities. Scatter diagram 1 shows sales of boat trip tickets and midday temperatures.

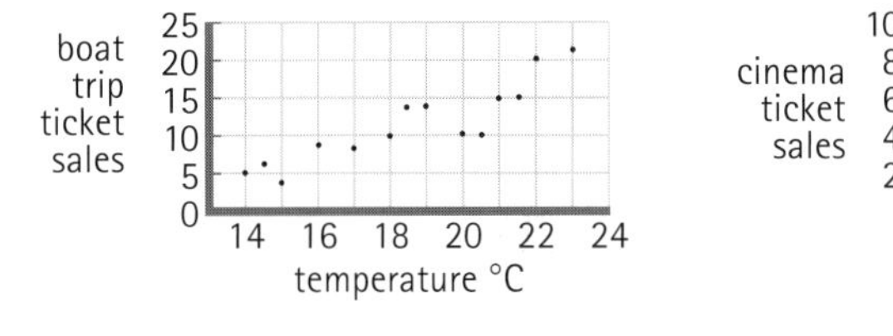

Scatter diagram 1

Scatter diagram 2

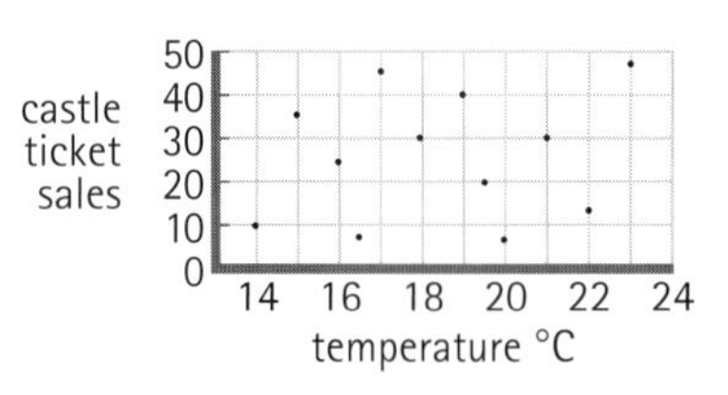

Scatter diagram 3

Scatter diagram 2 shows sales of cinema tickets and midday temperatures.

Scatter diagram 3 shows sales of castle tickets and midday temperatures

Scatter diagram 1 shows as the temperature increases, the sales of boat tickets increase. This is an example of positive correlation:

- if one variable increases as the other increases, it is called **positive correlation**

Scatter diagram 2 shows as the temperature increases, the sales of cinema tickets decrease. This is an example of negative correlation:

- if one variable increases as the other decreases, it is called **negative correlation**

Diagram 3 shows no correlation at all.

Practice questions

1) Annie and Derek go bowling. Their score sheets look like this.

Annie: number of skittles knocked over

2 4 2 1 8 0 1 2 10 0

Derek: number of skittles knocked over

6 9 6 1 6 5 5 7 2 1

a) What is Annie's:

(i) mean score

(ii) modal score

(iii) median score?

b) Whose scores have the greater range? Show your working.

2) Look at these two graphs.

Conversion graph for pounds £ to francs fr

Conversion graph for pounds £ to dollars $

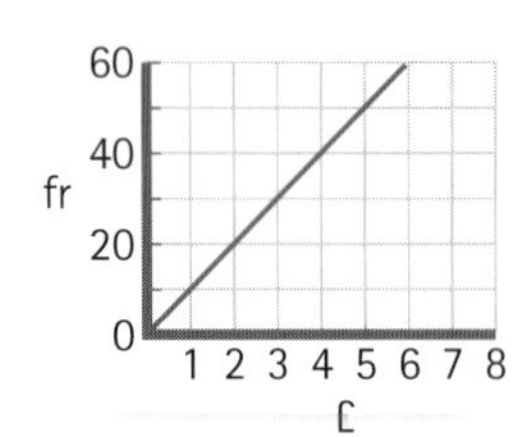

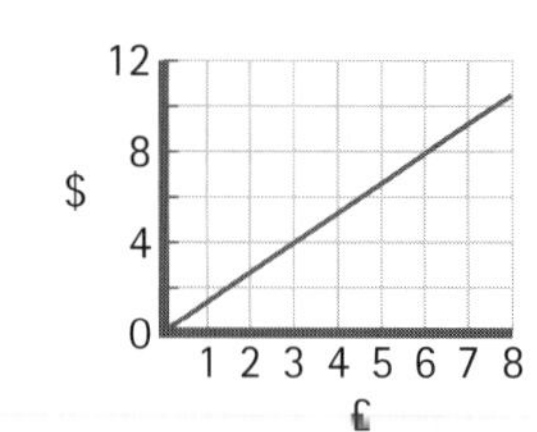

a) How many pounds would you get for 20 francs?

b) How many pounds would you get for 4 dollars?

c) Would you get more pounds for 50fr or $8?

Draw lines on the graphs to help you.

Probability

In this unit you will revise:

- understanding probability
- the probability scale
- calculating probability

Understanding probability

Probability is a measure of chance. Chance is sometimes described in words:

impossible unlikely very likely certain evens likely very unlikely

These words could be written on a scale:

impossible .... certain

• Write the rest of the words on the scale, from greatest to the least chance.

These words describe the chance of a particular outcome of an event happening.

An **event** is something that happens, such as a coin being flipped or a card being picked.

An **outcome** is the result: for a coin it is a head or a tail.

Examples

a) it is certain that this book will help you revise mathematics

b) it is very unlikely that someone in your family wins the Lottery

c) it is evens that if you toss a coin the outcome will be heads

Can you think of some more outcomes and events that could be described using these words?

The probability scale

Probability is the number assigned to the chance of a particular outcome of an event. Probability is measured as a fraction between 0 and 1.

0 is impossible $\frac{1}{2}$ is evens 1 is certain

So the probabilities for events given as examples above are:

a) this book will help you revise mathematics

0 ½ 1

b) you toss a coin and the outcome is heads

0 ½ 1

Calculating probability

The probability (or chance) of an outcome can be calculated by theory, or experiment and then using this formula:

$$P(\text{outcome}) = \frac{\text{number of favourable outcomes}}{\text{number of all possible outcomes}}$$

By theory:

Example 1

To find the probability of rolling a 3 or 5 on a die:

- favourable outcomes = 3 or 5 so there are 2 favourable outcomes
- all possible outcomes = 1, 2, 3, 4, 5 or 6 so there are 6 possible outcomes

Therefore P(rolling a 3 or a 5 on a die) = $\frac{2}{6} = \frac{1}{3}$
assuming that the die is fair, so each number is equally likely to be rolled.

Example 2

Find the probability of choosing a square number on a raffle ticket from a box of tickets numbered 1–50.

favourable outcomes = 1, 4, 9, 25, 36, 49 so there are 6 favourable outcomes

all possible outcomes = 1–50 so there are 50 possible outcomes

Therefore P(choosing a square number) = $\frac{6}{50} = \frac{3}{25}$
assuming that the raffle is fair, so each ticket is equally likely to be chosen.

There are nine tomatoes in a bag, seven are red and two are green. Each one has an equal chance of being selected.

- Write down the probability of selecting a green tomato.

By experiment:

Example

Estimate the probability of a slice of toast landing 'butter-side up' or 'butter-side down' when dropped on the floor.

Freda drops 25 slices of buttered toast from a tray. These are her results.

Butter-side up	Butter-side down
17	8

favourable outcome: butter-side up = 17
all possible outcomes = 25
so P(butter side up) = $\frac{17}{25}$

favourable outcome butter-side down = 8
all possible outcomes = 25
so P(butter side down) = $\frac{8}{25}$

Freda repeats her experiment twice more. These are her results.

Experiment 2

Butter-side up	Butter-side down
15	10

Experiment 3

Butter-side up	Butter-side down
18	7

Find P(butter side up) and P(butter side down) for each of these experiments.

You should have written:

P(butter side up) = $\frac{15}{25} = \frac{3}{5}$ P(butter side up) = $\frac{18}{25}$

P(butter side down) = $\frac{10}{25} = \frac{2}{5}$ P(butter side down) = $\frac{7}{25}$

Notice, Freda does not get exactly the same results each time, but her results are close.

Probability

Practice questions

1) Mrs Thompson has five children. Every morning she wakes them up for school. Some of them are good at getting up. Others are not so good! She does a survey over 10 days to show them her difficult task every morning.

Barney gets up on time one time out of ten.
Felix gets up on time five times out of ten.
Daisy gets up on time eight times out of ten.
Isaac gets up on time every morning.
Millie never gets up on time.

Write the children's names on the scale to show the probability of them getting up on time.

2) Mrs Goulbourn wants her class to work in groups on different mathematical activities. She gives each pupil a number 1–25.

1 2 3 4 5 6 7 8 9 10

11 12 13 14 15 16 17 18 19 20

21 22 23 24 25

Cross off each number as it is used.

a) All pupils with a number in the 7 times table work on activity 1.
What is the probability that a pupil's number is in the 7 times table?

b) All pupils with a 3 in their number work on activity 2.
What is the probability that a pupil has a 3 in their number?

c) All pupils with a number divisible by 6 work on activity 3.
What is the probability that a pupil's number is divisible by 6?

d) All pupils with a square number work on activity 4.
What is the probability that a pupil's number is a square number?

e) All pupils with a number in the 11 times table work on activity 5.
What is the probability that a pupil's number is in the 11 times table?

f) The rest of the pupils work on activity 6.
What is the probability that a pupil is in this group?

3) Miranda cannot decide whether to buy George Michael's new CD or Cher's new CD. She decides to toss a coin twice.

a) One outcome is heads then heads. What other three outcomes are there?
She decides:
2 heads or 2 tails means George Michael,
1 head and 1 tail means Cher.

b) Is Miranda's experiment fair?

c) What is the probability that Miranda buys Cher's new CD?

Mental maths

In this unit you will revise:

- preparing for a mental maths test
- practising mental maths with test questions

Preparing for the mental maths tests

The only way to prepare for a mental maths test is practice.

You will need to photocopy the mental maths tests answer sheets on pages 95 and 96 of this book.

Test 1 is a lower-tier test for those taking Tier 3–5 papers.

Test 2 is a higher tier test for those taking Tier 4–6 papers and above.

Each test should last approximately 20 minutes.

You must work out all your answers in your head and write them in the boxes. You are only allowed pens and pencils – **no other equipment.**

For some of the questions important information is already written down on the answer sheet.

Ask a friend or a parent to read the questions below to you, for either Test 1 or Test 2. They should read each question twice and then you should be given the time indicated to answer. The correct answers are given on page 94.

Mental maths test 1 questions (lower tier)

Please read this carefully.

For this first group of questions you'll have 5 seconds to work out each answer and write it down.

1. A room is 300 centimetres wide. How many metres is that?
2. Divide forty-eight by eight.
3. What number should you add to twenty-seven to make forty?
4. Look at the four numbers on your answer sheet. Find their sum.
5. Multiply six by six and then add three.
6. How many degrees are there in two right angles?
7. The diagram on your answer sheet shows part of a ruler. Write the missing number.

For the next group of questions you'll have 10 seconds to work out each answer and write it down.

8. The time is 13:25. How many minutes is this before two o'clock in the afternoon?
9. Simplify the expression on your answer sheet.
10. A square has sides of length 7cm. What is the perimeter of the square?
11. The bar chart shows the numbers of pets owned by pupils in a class. 9 pupils owned one particular type of pet. What type of pet was it?
12. Look at the thermometer on your answer sheet. What temperature is it?
13. 32% of the members of the hockey club are female. What percentage are male?
14. Larry bought a computer for £1000. He sold it three years later for half that price. How much did he sell his computer for?
15. 60.96 metres is the same length as 200 feet. How many metres is the same length as 100 feet?
16. A medium-sized city has a population of three million and twenty-one thousand. Write this number in figures.
17. Multiply eleven by sixty.
18. Look at the expression on your answer sheet. What is its value when a is eight and b is three?
19. A jacket cost £40. The price went up by 10%. What is the new price of the jacket?
20. What is a half add a quarter?

For the next group of questions you'll have 15 seconds to work out each answer and write it down.

21. A shop sells CDs at £10.99. How much would 5 CDs cost?
22. In a game of chance, balls numbered 1 to 20 are put into a bag. A ball is taken out of the bag at random. Draw an arrow on the scale to show the probability that the ball has an odd number on it.
23. In a survey people were asked, 'Do you have breakfast every day?' The diagram shows the result. Approximately what percentage of people said no?
24. There are thirty pupils in a class. There are twice as many girls as boys. How many girls are there?
25. You buy three items. Your answer sheet shows the cost of each one. How much change do you get from one pound?

26. Look at the diagrams on your answer sheet. Tick below the diagram that shows the net of a cube.
27. Look at the relationship on your answer sheet. If the first number became 4.12, what should the second number be?
28. Look at the shape inside the box. One of the diagrams beside it shows the shape rotated ninety degrees clockwise. Circle the correct diagram.
29. There are 180 passengers on a train. The pie chart shows which numbered carriages they are travelling in. About how many passengers are in carriage two?

Put your pen down. The test has finished.

Mental maths test 2 questions (higher tier)

Please read this carefully.

For this first group of questions you'll have 5 seconds to work out each answer and write it down.

1. Look at the numbers on your number sheet. Find a quarter of their total.
2. The probability that I get a five on a die is one sixth. What is the probability that I do not get a five?
3. How many millimetres make 11 centimetres?
4. Simplify the expression on your answer sheet.
5. What is 6 multiplied by 5 multiplied by 4?
6. Write the number 788.9 to the nearest 10.

For the next group of questions you'll have 10 seconds to work out each answer and write it down.

7. Look at the angle on your answer sheet. Estimate its size in degrees.
8. Look at the expression. Find its value when x equals 7.
9. What is 230 minus 72?
10. What is 10% of £320?
11. Look at the triangle on your answer sheet. What is its area?
12. Look at the diagram. Angle s is 93 degrees, how many degrees is angle t?
13. 25% of a number is 9. What is the number?
14. The fraction of boys in a class is $\frac{3}{5}$. What fraction of the class are girls?
15. n stands for a number. Write an expression for the following: multiply n by 4, then add 7 to the result.

16. Two weeks ago the height of my sunflower was 3.4cm. Now the height of my sunflower is 12.1 cm. How many centimetres has my sunflower grown?

17. My heart beats 70 times every minute. How many times will it beat in one hour?

18. Work out 0.04 multiplied by 6.

19. Look at the equation on your answer sheet. Use it to work out the value of $2x$.

For the next group of questions you'll have 15 seconds to work out each answer and write it down.

20. Look at the cube. How many small cubes would be needed to make a cube of dimensions twice the size of the dimensions of this cube?

21. Look at these numbers. Circle the smallest number.

22. What is one third of six hundred thousand? Write your answer in figures.

23. Look at the lengths on your answer sheet. Put a tick below the median length.

23. Look at the calculation on your answer sheet. Use it to help you work out the answer to 23 multiplied by 41.

25. A corn circle has circumference 21 metres. Approximately what is its diameter?

26. You get 180 francs for 20 pounds. How many francs do your get for 100 pounds?

27. Write an approximate answer to the calculation.

28. Look at the expression on your answer sheet. a and b are consecutive integers. What are the values of a and b.

29. Divide £66 equally between twelve people. How much money does each person get?

30. Look at the pairs of numbers on your answer sheet. Between which pair of numbers does the square root of 110 lie?

Put your pen down. The test has finished.

Answers to example questions

Page 7

- two thousand, nine hundred and thirty-seven =2937.
- 900 is 900 units; 900 is 90 tens; 900 is 9 hundreds

Page 9

- 2000 x 70 =140 000 (2 x 7 = 14 and 1000 x 10 = 10 000)
- 900 ÷ 30 = 30 (900 ÷ 10 = 90 and 90 ÷ 3 = 30)

Page 11

- £0.04 (0.04 is 4 hundredths)
- £28.09
- 2.43 + 34.732 + 125 = 162.162

Page 12

- a) 36; b) 6394.1
- a) 3.29; b) 0.5114
- a) 22.452; b) 105.4

Page 14

- $\frac{3}{7}$

Page 15

- $\frac{1}{3} = \frac{5}{15}$ and $\frac{3}{5} = \frac{9}{15}$ so $\frac{5}{15} + \frac{9}{15} = \frac{14}{15}$

Page 16

- $\frac{2}{3}$ of 27 = 18 as 27 is split into 3 equal parts (thirds), so

  $\frac{1}{3}$ of 27 is 27 ÷ 3 = 9; 2 of those equal parts make the

  fraction $\frac{2}{3}$, so $\frac{2}{3}$ of 27 is 2 x 9 = 18

Page 18

- a) £16.95 (56.5 x £30 = £1695 and £1695 ÷ 100 = £16.95);
  b) 11.16 g (9 x 124 g = 1116 g and 1116 g ÷ 100 = 11.16 g)

Page 20

- a) −10 is larger than −24 b) −183 is smaller than −83

Page 21

- −5 + 10 − 2 − 7 =−4, because, starting from zero, you count −5, then count +10 (taking you to +5), then count −2 (taking you to +3), then count −7 (taking you to −4)

Page 22

- a) 13; b) 10

Page 23

- a) 300 because 251 is closer to 300 than 200; b) 2880 because 2883 is closer to 2880 than 2890

Page 24

- 8 mugs = 8 x 8 = 64 cm; 11 mugs = 11 x 8 = 88 cm

Page 26

- $p + m = 2 + 0.45 = £2.45$

Page 27

- a) g x 4 or 4g; b) 16 + 11 = 27; c) $h \div 5$ or $\frac{h}{5}$
- a) 34 b) $\frac{t}{3} + 9$

Page 28

- a) ÷ 4; b) x 5 − 11
- a) $15u + 2u = 17u$; b) $v + 6v = 7v$

Page 29

- a) $7m - 6n$; b) $15y - 13$

Page 32

- a) 19°C because C = (68 − 30) ÷ 2 = 38 ÷ 2 = 19°C;
  b) 28°C because C = (86 − 30) ÷ 2 = 56 ÷ 2 = 28°C;
  c) 21°C because C = (72 − 30) ÷ 2 = 42 ÷ 2 = 21°C
- triangle: $\frac{1}{2}$ x 5 x 1.2 = 3 cm^2
- cuboid: 2 x 2.4 x 6 = 28.8 cm^3

Page 34

- a) rule: add 3, the next number is 15 because 12 + 3 =15; b) rule: multiply by 2, the next number is 16 because 8 x 2 =16

Page 37

- a) x 8; b) − 7; c) + 1 ÷ 4

Page 38

- a) y = 6 because y → x 8 → 48 applying the inverse rule: 48 → ÷ 8 → 6, so y = 6;
  b) r = 2 because r → x 9 → − 12 → 6 applying the inverse rule: 6 → + 12 → ÷ 9= 2 so r = 2;
  c) a = 5 because a → + 23 → ÷ 4 → 7. Now applying the inverse rule: 7 → x 4 → − 23 → 5 so a = 5

Page 40

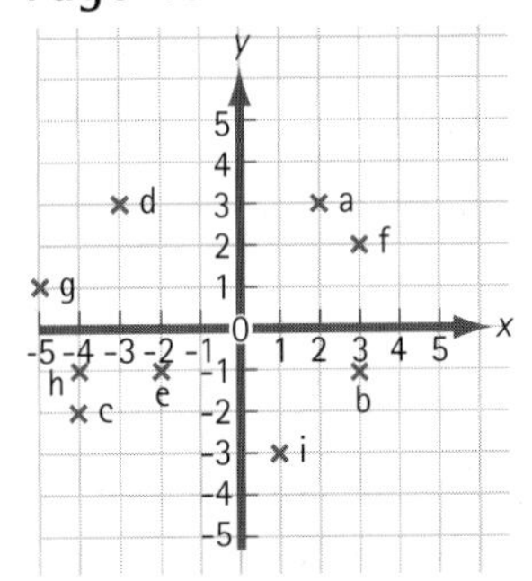

Page 41

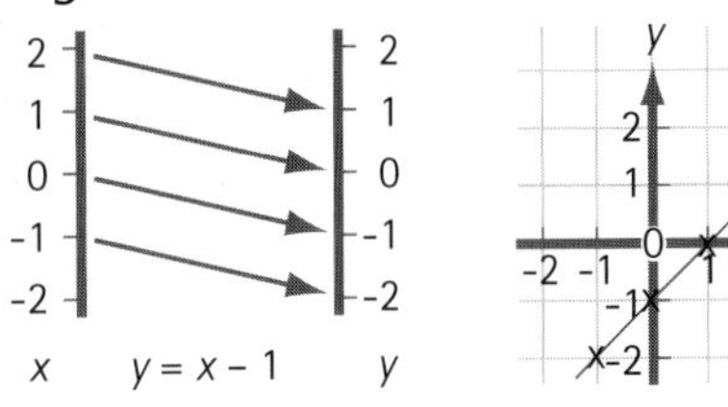

Page 42

- a) x = 8 (the x-coordinate of every point on the line through b and c is 8);
  b) y = x − 8 (the line through points c and d is parallel to the line through points b and e that has the equation y = x. If you extend the line through c and d it crosses the y-axis at y = −8.

Page 43

- a) is obtuse; b) is reflex; c) is obtuse; d) is acute.

Page 44

- a) 130°; b) 60°
- a) b)

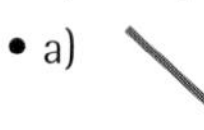

Page 45

- a) 39° because 90° – 51° = 39°;
  b) 249° because 360° – 111° = 249°;
  c) 153° because 360° – 207° = 153°;
  d) 134° because 180° – 46° = 134°

Page 49

- a) 127° (corresponding angles);
  b) 127° (vertically opposite angles);
  c) 53° (180° – 127° = 53°) (angles on a straight line);
  d) 139° (180° – 41° = 139°) (interior angles);
  e) 139° (alternate angles);
  f) 139° (vertically opposite angles)

Page 50

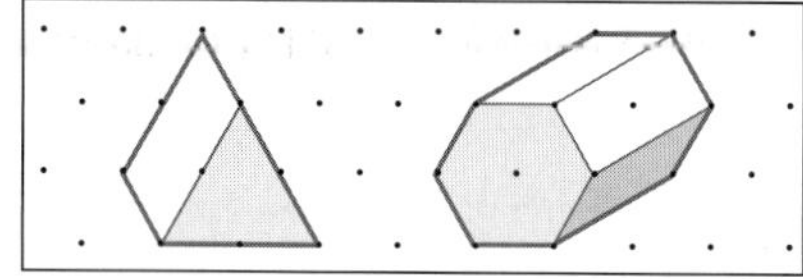

Page 51

- top elevation, rectangle; front elevation, triangle; side elevation, rectangle

Page 55

- sum of the interior angles of a parallelogram = 2 x 180° = 360°; sum of the interior angles of a pentagon = 3 x 180° = 540°; sum of the interior angles of a rhombus = 2 x 180° = 360°

Page 56

- x = 95° because 180° – 85° = 95°; y = 83° because 360° – 92° – 95° – 90° = 83°; z = 97° because 180° – 83° = 97°

Page 57

- a) b) c) d

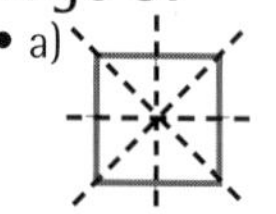

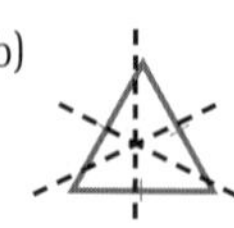

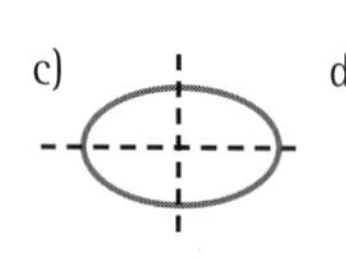

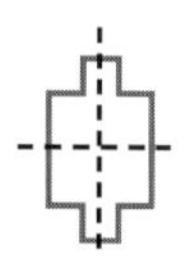

Page 58

- a) order 2; b) order 5 c) order 4 d) order 8

Page 59

- (b), (c) and (f) are congruent because (c) is (b) rotated and then reflected, and (f) is (b) rotated.
- [3 squares left, 1 square down]

Page 61

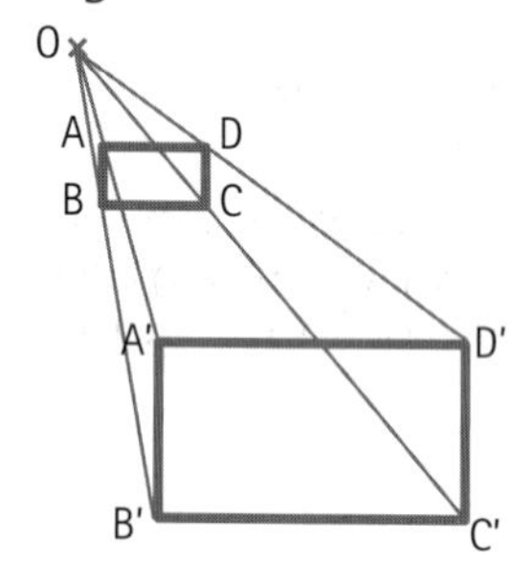

Page 63

- each division is 5°C and the temperature is 25°C
- a) 37.8 cm (mm → cm, small → large, so ÷ and 10 mm = 1 cm, so 378 ÷ 10 = 37.8 cm);
  b) 48 pts (gallons → pints, large → small, so x and 8 pt = 1 gallon, so 6 x 8 = 48 pts)

Page 69

- 7 units squared (4 x □ and 6 x ◺ = 4 x □ and 3 x □)

Page 71

- a) area = l x l = 8 x 8 = 64 cm²; b) area = $\frac{1}{2}(a + b)h$ = $\frac{1}{2}(7 + 5) \times 4$ = $\frac{1}{2}$ (12 x 4) = $\frac{48}{2}$ = 24 mm²; c) area = $\pi r^2 = \pi \times 2 \times 2$ = 3.14 x 2 x 2 = 12.56 m²

Page 77

- a) $\frac{1}{4}$ of 120 = $\frac{1}{4}$ x 120 = 30 pupils travel by bus;
  b) $\frac{1}{6}$ of 120 = $\frac{1}{6}$ x 120 = 20 pupils walk

Page 80

- a) The median is 32.5 seconds because it is the middle time in size order. The mode is 30 seconds because 30 seconds is the most common time. The mean is 33 seconds because the numbers added together = 528 and there are 16 times altogether, so the mean = 528 ÷ 16 = 33 second.
  b) Mary's time range is 39 – 33 = 6 seconds; Nigel's time range = 32 – 30 = 2 seconds; Amy's time range = 35 – 26 = 9 seconds; Oliver's time range = 40 – 32 = 8 second. So the coach puts the team in the order Nigel, Mary, Oliver, Amy.
- From the pie chart, $\frac{1}{4}$ of Class M2 play volleyball, $\frac{1}{4}$ of 24 = 6 pupils. From the bar chart, 7 pupils in Class N2 play brass instruments. So there are fewer volleyball players in Class M2 than brass players in Class N2.

Page 83

impossible very unlikely unlikely evens likely very likely certain

Page 84

- $\frac{2}{9}$ favourable outcome = green tomato so there are 2 favourable outcomes all possible outcomes = all tomatoes so there are 9 possible outcomes.

Answers to practice questions

Number

Whole numbers

1) a) Any 4 digit number beginning 10.. or 12.. b) 984; c) 0

2) 100 x 4, 8000 ÷ 20, 40 tens, 20 x 20, 4000 ÷ 10, 1200 ÷ 3, 8 x 50

Decimals

1) a) 10.2; b) 9.1; c) 1.03; d) 9.09; e) 9.10; f) 10.3; g) 100; h) 9.8; i) 9.11

2) a) £1.03; b) £8.53; c) £10.30; d) £10.14

Fractions

1) a) $\frac{3}{15}$ or $\frac{1}{5}$; b) $\frac{2}{5}$; c) E; d) A; e) B and C; f) D

2) a) £15; b) £48; c) £50; d) £7

Percentages

1) A and E, B and H, C and F, D and G

2) a) 50% of £60 is £30, 25% of £60 is £15, so 75% of £60 is £45; b) £6; c) 10% = £6, 5% = £3, 2.5% = £1.50, Carlo's sister gets £1.50

3) a) 24.8 metres; b) 103.6 grams; c) £82.72

4) 55% boys

Negative numbers

1) a) −15°C, −8°C, −1°C, 0°C, 7°C, 12°C; b) −15°C and −8°C

2) a)-20 and +5 b) -4, -2 and 0

c) + 2 − 10 = −8; d) − 3 + 2 − 9 = −10; e) + 8 − 8 = 0; f) + 4 − 11 = − 7

Rounding and approximating

1) a) 2000; b) 300; c) 610; d) 28; e) 4400

2) a) 30 x 40 = 1200; b) $\frac{300-60}{6}$ = 40

3) 50

Unitary method and ratio

1) £140 2) £57.20 3) a) 9 pints; b) 15 pints

Algebra

Simple formulae

1) a) 9; b) 8; c) y; d) $3z - 8$; e) 3

2) a) $4y$; b) $12y$; c) 5 cm

3) a) $x - 2$ b) $3x$; c) 15; d) 39

4) a) $7 + 2b$; b) $b + 1$

Using formulae

1) a) 15 kilometres per hour; b) 12 kilometres per hour

2) a) £3428; b) £2140; c) £1585

3) 18 cm²

Number patterns

1) a) 66, 55, 44, 33, 22, 11 the rule is subtract 11;
b) 100 000 → 10 000 → 1000 → 100 →10; rule is divide by 10

2) a) 4; b) 2; c) 14 white tiles, 11 grey tiles;

d)

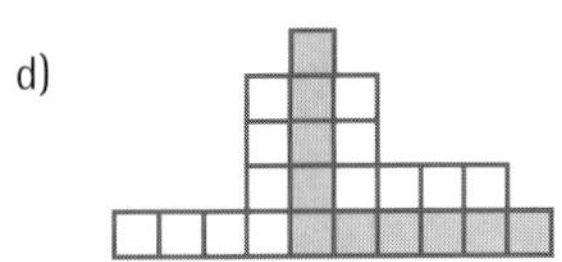

e) 6n + 1 For 100th pattern 601 tiles will be needed.

Equations

1) a) − 5; b) − 1 ÷ 25; c) ÷ 5; d) x 25; e) ÷ 25 − 1

2) a) $n = 5$; b) $x = 6$; c) $p = 2$

3) a) $3n + 1 = 46$; b) 15

Coordinates, mappings and line graphs

1) a) (2 , 2) (6 , 3); b) (14 , 5)

Shape and space

Angles

1) a = 12°; b = 151° 2) Check the angles are correct

3) a = 34° (quarter turn); b = 230° (full turn); c = 43° (quarter turn); d = 56° (half turn); e = 8° (half turn); f = 311° (full turn); g = 29° (quarter turn)

Parallels and angles

1) p = 141° (interior angles)

2) q = 129° (corresponding angles)

3) r = 89° (alternate angles)

4) s = 37° (vertically opposite angles)

4) t = 22° (vertically opposite, then corresponding angles)

2D shapes and 3D solids

1) or perhaps

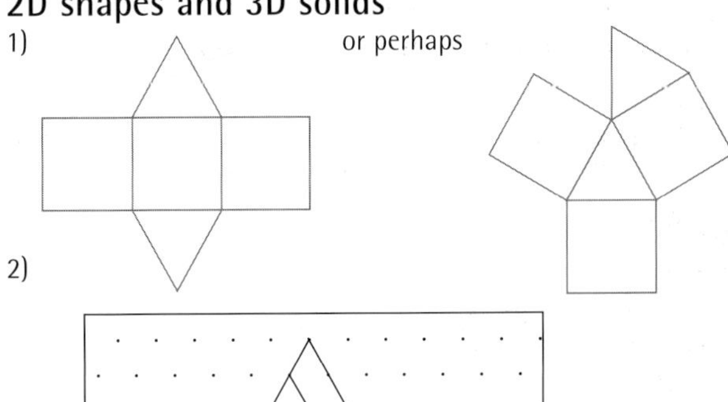

2)

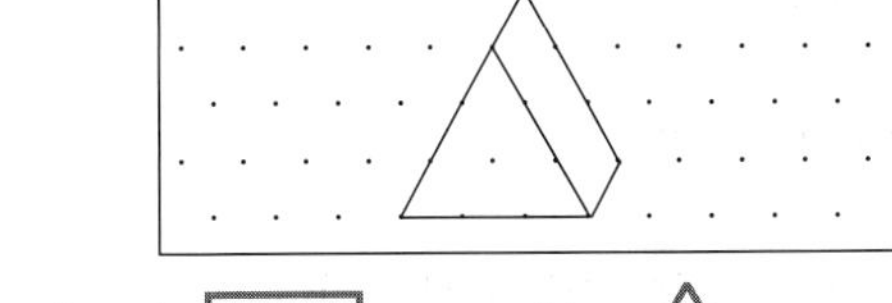

3) a)

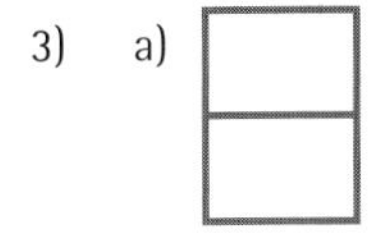

b)

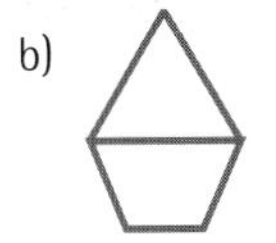

c)

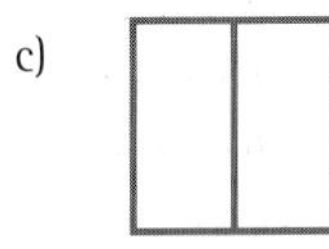

Polygons

1) a) rectangle; b) trapezium; c) rhombus; d) triangle;
e) irregular hexagon; f) parallelogram; g) kite

2) a = 130° (angle sum of a trapezium); b = 50° (angle sum of a trapezium); c = 30° (angles on a straight line); d = 60° (angle sum of a triangle); e = 155° (angle sum of a hexagon); f = 25° (angles on a straight line)

Symmetry and congruence

1) a)

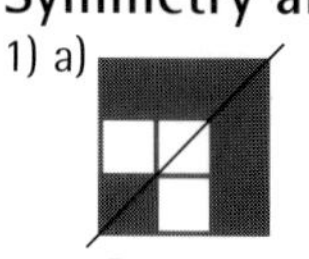

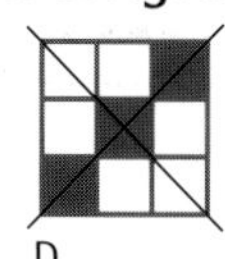

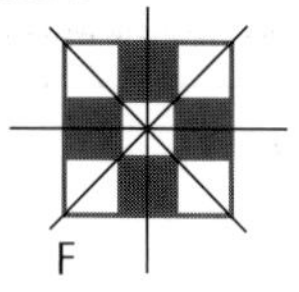

B D F

b) A has rotational symmetry of order 2. E has no rotational symmetry or order 1. F has rotational symmetry of order 4.

c) The congruent pairs of tiles are: A and D, B and E, C and G

2) a) no rotation; b) 3 squares left, 2 squares up.

Enlargment and translation

1)

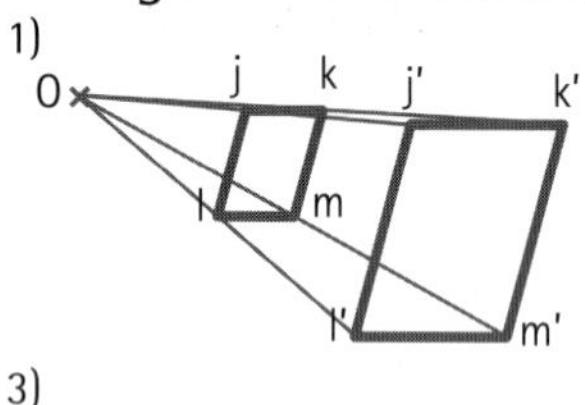

2)

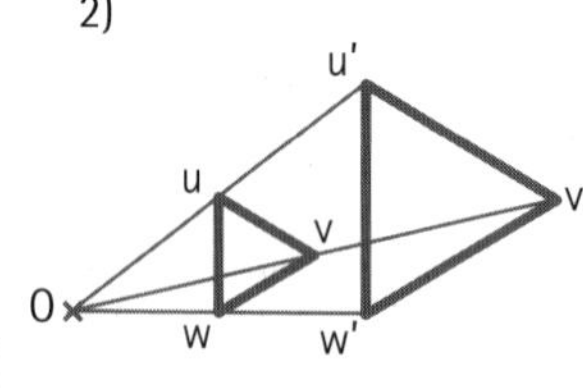

3)

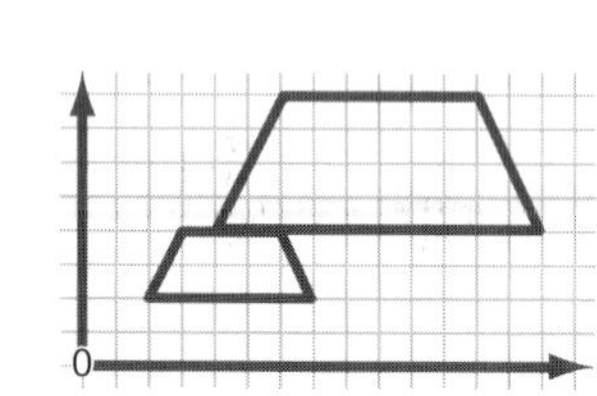

4)

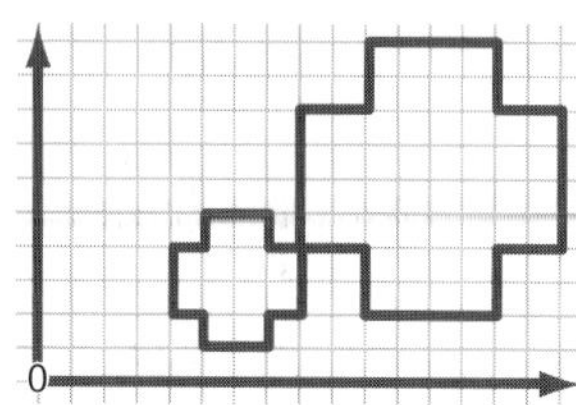

Measures

Units of measure

1) a) approx. 4.5 cm; b) approx. 750 ml; c) approx. –5°C

2) a) 10 in; b) 0.95 m; c) 330 mm; d) 135 cm; 3) Sarah runs 3 miles = 4.8 km; Mark runs 3.125 miles = 5 km; Mark runs further.

Perimeter

1) a) 9 cm b) 17 cm

2) a) (i) 15.7 cm^2; (ii) approx. 15 cm^2 b) (i) 9.42 cm^2; (ii) approx. 9 cm^2; c) (i) 34.54 cm^2; (ii) approx. 33 cm^2

Area

1) 6 units squared
2) 8 units squared
3) 13 units squared
4) 15 cm^2
5) 7 cm^2
6) 16 cm^2
7) 12 cm^2
8) 9 cm^2
9) 113.04 mm^2
10) 120 mm^2
11) 78.5 cm^2
12) 9x

Volume

1) 45 units cubed; 2) 30 units cubed; 3) 8 units cubed;
4) 24 cm^3; 5) 84 cm^3; 6) 36 cm^3

Handling data

Collecting and interpreting data

1) WWF (World Wildlife Fund) 2) $\frac{1}{4}$ 3) 27 pupils

4) 7 sponsored events 5) sales 6) £150 7) £1000

Comparing distributions

1) a) (i) 3 (ii) 2 (iii) 2

b) Annie's score range: 10 – 0 = 10; Derek's score range: 9 – 1 = 8; Annie's scores have the greater range

2) a) £2; b) £3; c) 50fr = £5, $8 = £6, you would get more pounds for $8

Probability

1)

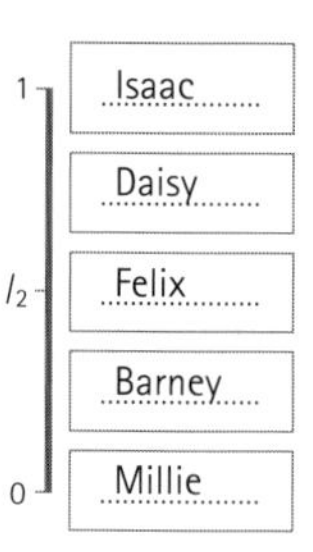

2) a) $\frac{3}{25}$; b) $\frac{3}{25}$; c) $\frac{4}{25}$; d) $\frac{5}{25}$ or $\frac{1}{25}$; e) $\frac{2}{25}$; f) $\frac{8}{25}$

3) a) tails then tails, heads then tails, tails then heads; b) Yes; c) $\frac{1}{2}$

Mental maths

Mental maths Test 1

1) 3 metres 2) 6 3) 13 4) 12 5) 39
6) 180* 7) 7.5 8) 35 minutes 9) 4s
10) 28 cm 11) cat 12) –1°C 13) 68% 14) £500
15) 30.48 m 16) 3 021 000 17) 660 18) 37
19) £44 20) $\frac{3}{4}$ 21) £54.95
22)

0 ½ 1

23) 25% 24) 20 25) 30p
26) You should have ticked:

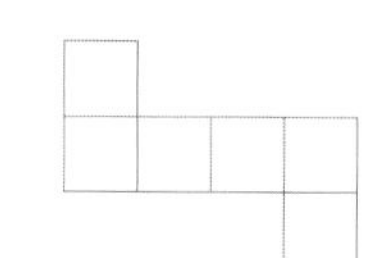

27) 16.81
28) You should have circled:

29) about 30

Mental maths Test 2

1) 5 2) $\frac{5}{6}$ 3) 110 millimetres 4) 5x
5) 120 6) 790 7) 141° 8) 17 9) 158
10) £32 11) 4.2 cm^2 12) 87° 13) 36
14) $\frac{2}{5}$ 15) 4n + 7 16) 8.7cm 17) 4200
18) 0.24 19) 44 20) 64
21) Circled –2 22) 200 000 23) 8m
24) 943 25) 7 metres 26) 900 francs
27) 40 28) 10 and 11 29) £5.50
30) 10 and 11

Mental Maths test 1 answer sheet

Time: 5 seconds

1 m

2

3 27 + = 40

4 1.9 4.1 4.1 1.9

5

6 degrees

7

6 6.5 7 ?

Time: 10 seconds

8 minutes

9 s + s + s + s

10 cm

11

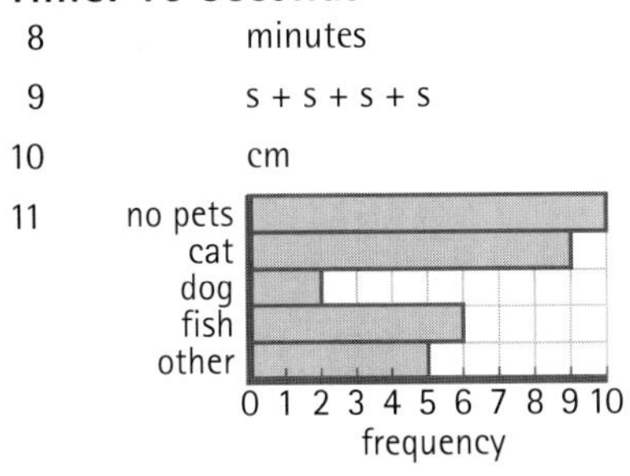

12 °C

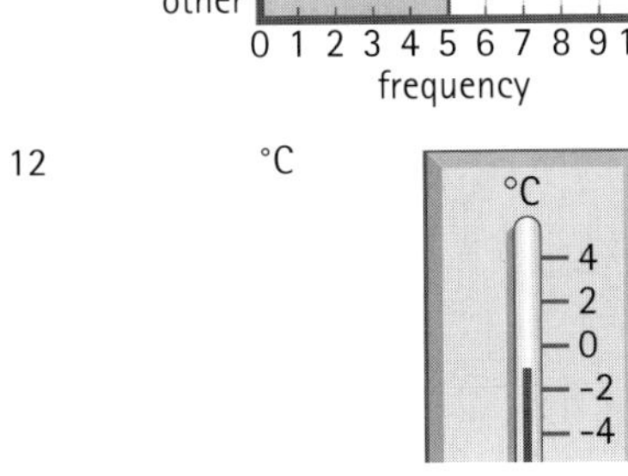

13 %

14 £

15 60.96 m = 200 feet

... m = 100 feet

16

17 11 x 60 =

18 $5a - b$

19 £

20

Time: 15 seconds

21 £

22

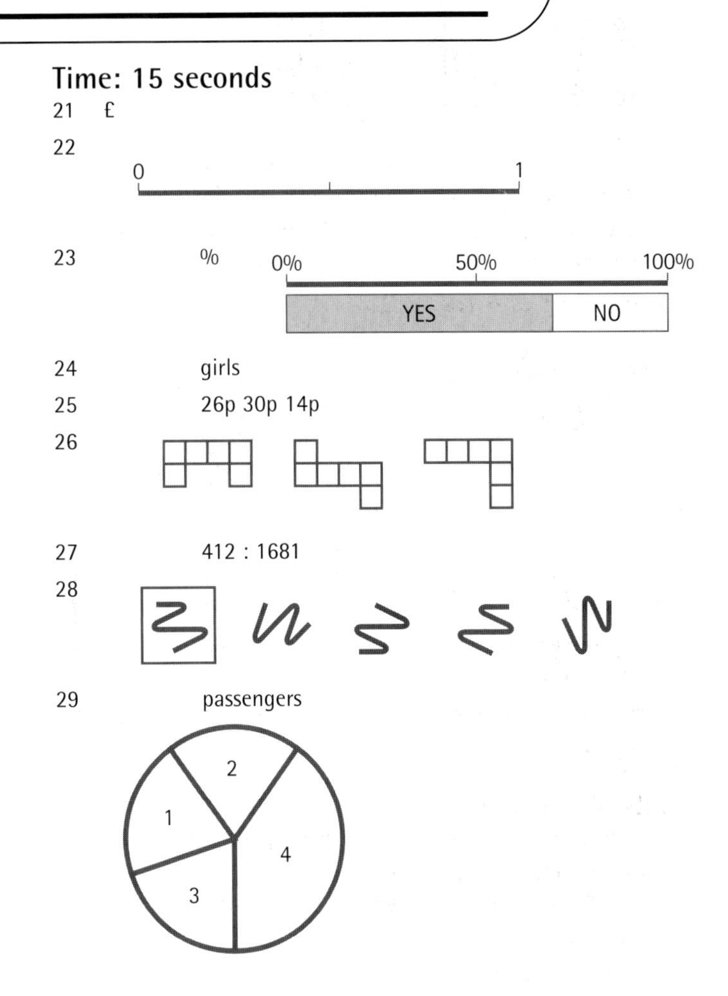

23 %

24 girls

25 26p 30p 14p

26

27 412 : 1681

28

29 passengers

Mental Maths test 2 answer sheet

Time: 5 seconds

1 10 6 4

2

3 mm

4 $13x - 8x$

5 6 5 4

6

Time: 10 seconds

7 ...°

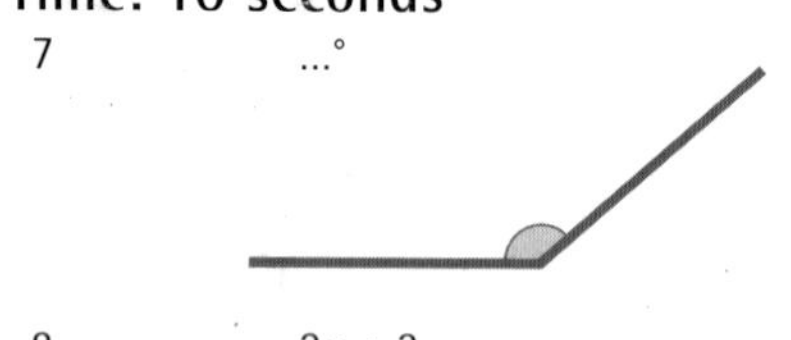

8 $2x + 3$

9 $230 - 72 =$

10 £

11 ... cm^2

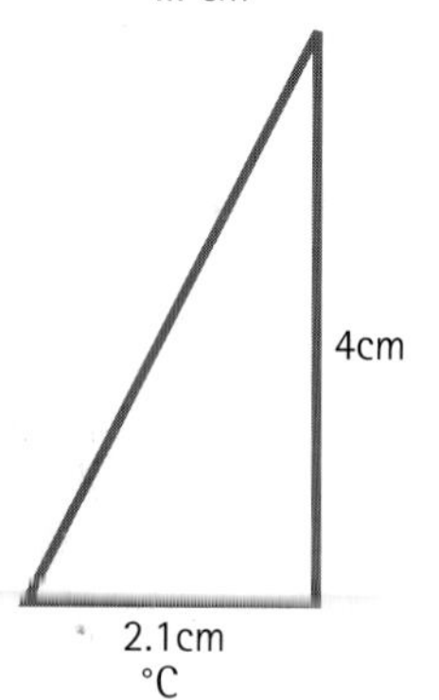

12 °C

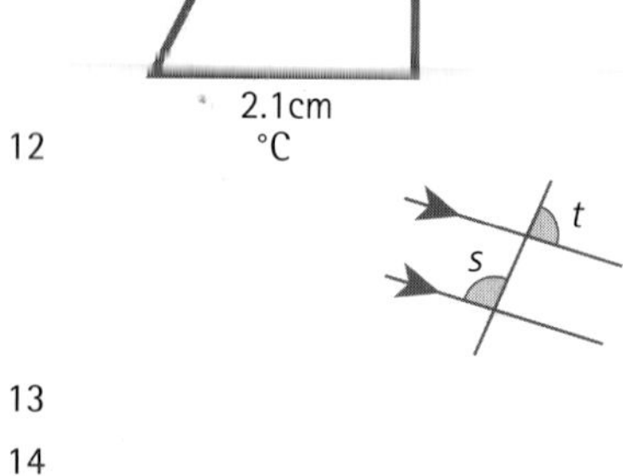

13

14

15

16 cm

17

18 0.04 6

19 $x - 3 = 19$

Time: 15 seconds

20

21 −2, −1.31, −1.5

22

23 13 m 8 m 7.5 m 1000 m 7.5 m

24 $23 \times 4 = 92$

23×41

25 m

26 francs

27 $\frac{103.9 + 21.7}{2.9}$

28 $a + b = 21$

29 £

30 10 and 11

... and ... 11 and 12

12 and 13